A
MATHEMATICAL TREATISE
ON
VIBRATIONS IN RAILWAY BRIDGES

A
MATHEMATICAL TREATISE
ON
VIBRATIONS IN RAILWAY BRIDGES

BY

C. E. INGLIS
M.A., F.R.S., M.Inst.C.E.

*Professor of Mechanism and Applied Mechanics
and Fellow of King's College in the
University of Cambridge*

CAMBRIDGE
AT THE UNIVERSITY PRESS
1934

CAMBRIDGE
UNIVERSITY PRESS

University Printing House, Cambridge CB2 8BS, United Kingdom

Cambridge University Press is part of the University of Cambridge.

It furthers the University's mission by disseminating knowledge in the pursuit of
education, learning and research at the highest international levels of excellence.

www.cambridge.org
Information on this title: www.cambridge.org/9781107536524

© Cambridge University Press 1934

First published 1934
First paperback edition 2015

A catalogue record for this publication is available from the British Library

ISBN 978-1-107-53652-4 Paperback

Cambridge University Press has no responsibility for the persistence or accuracy
of URLs for external or third-party internet websites referred to in this publication,
and does not guarantee that any content on such websites is, or will remain,
accurate or appropriate.

CONTENTS

INTRODUCTION

The object of the following treatise is the study of the dynamic effects in railway bridges, produced by the action of locomotives and other moving loads, and the determination of simple, yet scientific formulae, whereby these effects can be predicted with a reasonable degree of accuracy.

For the calculation of stresses due to statical loads, mathematical analysis has reached a stage of evolution which hardly admits of much further advance; but, on its dynamical side, the science of bridge design cannot be viewed with equal satisfaction.

This stage of arrested development is largely due to the blighting influence of impact factors, which prescribe that the dynamical effect of a moving load shall be taken into account by multiplying it by a factor depending upon the span of the bridge, and usually on nothing else. The simplicity of this process, which relieves bridge designers from intellectual effort, has ensured its popularity in the past, but the imposition of these purely empirical factors, which gain their simplicity only at the cost of ignoring all the considerations of real importance, has most effectively discouraged the enterprise bridge engineers might otherwise have shewn in developing their science to deal intelligently with dynamical stresses. Typical of these impact factors, and the one which perhaps has enjoyed the greatest vogue, is the Pencoyd formula $1 + \dfrac{300}{300 + l}$, where l is the span of the bridge measured in feet.

For short spans this impact factor imposes an addition of 100 per cent. to the live load, and this increment decreases as the span increases, so that when the span is 300 feet, the addition called for is reduced to 50 per cent.

In 1920, as the result of some experiments carried out by the Ministry of Transport, the Pencoyd formula was recast and reimposed in the form $1 + \dfrac{120}{90 + \dfrac{3l}{2}}$, while in other countries, impact factors, expressed as functions of the span, have come and gone with a bewildering rapidity, which is only to be expected from formulae which stand on no permanent logical or scientific foundation.

But to heap ridicule and contempt on these empirical formulae would be an act of base ingratitude, and, though one may cherish the hope that their reign is coming to an end, it should always be remembered to their credit that in the past they have performed good service in forcing engineers to build their bridges with such an ample margin of strength that in most cases they have been able to stand up to the increased loads demanded by modern requirements.

Purely empirical formulae, however, can have no permanent place in the science of engineering; sooner or later they must give way to formulae based upon scientific principles, formulae which bear the hall-mark of truth, in that they embody all the separate characteristics which theory, in conjunction with experiment, indicate must be taken into account.

The first really comprehensive attempt to put the problem of bridge impact on a scientific basis was made by the Bridge Stress Committee which was established under the Chairmanship of Sir Alfred Ewing in March 1923 and published its Report in October 1928. Previously, although a great mass of experimental data had been accumulated, in default of any underlying theory sufficiently comprehensive to explain the phenomena observed, the inferences which could be drawn from this welter of disconnected experimental results were disappointingly meagre. Experimental observations, uncontrolled by mathematical analysis, seldom lead to conclusions of permanent value, and this problem of bridge impact, like all great problems in engineering science, calls for a close cooperation between mathematical analysis and practical experiment. Mathematical analysis is required to indicate the lines along which the experiments should proceed, and experiment, in its turn, is necessary to check the validity of theoretical predictions and to prevent mathematics running off the scent and barking, so to speak, up the wrong tree.

As a Member of the Bridge Stress Committee, the writer was mainly engaged in evolving a satisfactory system of mathematical analysis, moulding it into shape until it fitted the facts, and pruning away unnecessary excrescences until it became a tool of reasonable simplicity and practical utility. Without the guidance of experimental results to point the way at every cross-road and to check the validity of analytical predictions, this task would have been quite impossible and, deprived of these experimental sign-posts, any mathematician, no matter how astute he might be, would inevitably lose his way in the labyrinth of side-tracks which confront the traveller in this comparatively unexplored territory.

Some of the analytical methods outlined in the Bridge Stress Report reappear in greater detail in the following Chapters; but the ground has been more thoroughly explored and the trail has been followed until definite formulae emerge, which are as simple as can be expected, in view of the numerous characteristics which insist on being taken into account. These formulae, though obtained by methods of approximation, yield results which can hardly be improved upon by the fullest possible system of mathematical analysis, and this is all the more gratifying because the full mathematical treatment, in its turn, is found to predict dynamical effects which are confirmed by experiment in a manner which leaves little to be desired.

The remainder of this introduction will now be devoted to a preliminary survey of the whole problem and an indication in broad outline of some of the more salient mechanical and analytical principles involved.

In this preface, which might be more correctly described as a summary, conclusions will be stated without proof, but references are given to the particular chapters and diagrams where the justification of these statements is to be found.

The vibrations set up by a train as it crosses a bridge are mainly caused by the "hammer-blows" due to the balance-weights which are attached to the driving-wheels of a locomotive for the purpose of minimizing the inertia effects of its reciprocating parts. Even when hammer-blows are absent, as in the case of electric locomotives, some vibration is inevitable, owing to the fact that the moving load is more or less suddenly applied, but oscillations of this character are very short lived and relatively insignificant in magnitude. Vibrations may also be caused by rail-joints, and the lurching consequent on track irregularities. Impact effects of this capricious character are obviously quite incalculable and, except possibly in the case of very short-span bridges, they are quite overwhelmed by the much larger oscillations due to hammer-blows. These latter oscillations, and the dynamical forces consequent thereon, come within the scope of exact mathematical analysis and, given the necessary bridge and locomotive characteristics, can be computed with a precision which is quite sufficient for all practical purposes.

Although it is recognized that empirical allowance must be made for the impact effects due to track irregularities, the main objective of this treatise is the analysis of the oscillations due to hammer-blows and the evolution of formulae for computing dynamic deflections and the bending-moments and shearing-forces induced thereby.

Certain general ideas concerning the behaviour of a bridge under the influence of repeated hammer-blows can be acquired by considering the somewhat analogous case of a spring-supported load acted upon by a periodic alternating force. Imagine, for instance, a mass suspended from a vertical helical spring. If displaced vertically above or below its position of statical equilibrium and then released, the mass will perform vertical oscillations having a definite frequency depending on the strength of the spring and the mass it supports.

Let n_0 denote this frequency, measured in oscillations per sec., and let e be the extension of the spring when a steady vertical force P is applied to the suspended mass. Then, under the influence of a periodic force alternating N times per sec., between the limits $\pm P$, the semi-amplitude of the oscillations, after the motion has become steady, is

$$e\,\frac{1}{1-\dfrac{N^2}{n_0{}^2}}.$$

To be more precise, if the alternating applied force is given by $P \sin 2\pi Nt$, the elongation of the spring at any instant is

$$e\,\frac{\sin 2\pi Nt}{1-\dfrac{N^2}{n_0{}^2}}.$$

Thus, when N is less than n_0, the oscillations are in phase with the applied force, and when the force pulls downward the spring elongates. If N is greater than n_0, the oscillations are anti-phased to the applied force, and while the force is acting downwards the suspended mass is moving upwards, and vice versa. When $N = n_0$, a condition of resonance or synchronism is established and, in the absence of damping, the oscillations will mount up to an infinite extent.

In this case, though not immediately apparent from the formula given above, the oscillations are in quadrature with the applied force, that is to say, when the force is zero the displacement of the suspended mass from its mid-position is a maximum.

In this behaviour of the spring-supported mass there is something almost human; it objects to being rushed. If coaxed gently and not hurried too much, it responds with perfect docility; but if urged to bestir itself at more than its normal gait, it exhibits a mulish perversity of disposition. Such movement as it makes under this compulsion is always in a retrograde direction, and the more it is rushed the less it condescends to move. On the other hand, if it is stimulated with its own natural inborn frequency, it plays up with an exuberance of spirit which may be very embarrassing.

But as an explanation of the behaviour of a railway bridge under the influence of the hammer-blows of a locomotive this analogy with a loaded helical spring is very far from being complete.

A bridge certainly has some resemblance to a spring, but, unlike a loaded helical spring, it possesses an infinite variety of modes of vibration, each of which has its own natural frequency, depending upon the mass and stiffness of the bridge and the loads it may be carrying.

The natural frequencies and normal modes of vibration of a bridge, both in its loaded and unloaded state, are dealt with in Chapter II, and it appears that the fundamental mode of vibration, that is the one having the lowest frequency and the only one which is likely to be stimulated into activity by the hammer-blows of a locomotive, approximates very closely to a simple sine curve having the equation

$$y = A_1 \sin \frac{\pi x}{l},$$

where y is the vertical deflection at a section distant x from an end of the bridge, and l is the length of the span.

The natural frequency, n_0, of this fundamental mode of vibration is given approximately by the formula

$$2\pi n_0 = \frac{\pi^2}{l^2} \sqrt{\frac{E I l}{M_G}},$$

where M_G is the total mass of the bridge, and I is the relevant moment of inertia of its mid-cross-section.

If the bridge supports a mass M at a section distant a from an end, the fundamental frequency is reduced, and the formula for determining its value, n_1, is

$$2\pi n_1 = \frac{\pi^2}{l^2} \sqrt{\frac{E I l}{M_G + 2M \sin^2 \frac{\pi a}{l}}}.$$

Other normal modes of vibration exist, approximating to the forms

$$y = A_2 \sin \frac{2\pi x}{l}, \quad y = A_3 \sin \frac{3\pi x}{l}, \quad \text{etc.,}$$

but the natural frequencies of these are so high that their influence on the oscillations induced in railway bridges is quite negligible.

Owing to practical limitations, the frequency of locomotive hammer-blows can never by any chance attain these high values and a frequency of 6 oscillations per sec. is broadly speaking the upper limit. Consequently, even when the span is as great as 300 feet, and the locomotive is travelling at its highest possible speed, synchronism with the higher modes of vibration of a railway bridge can never be achieved, and the only form of resonance possible is that which occurs when the hammer-blows synchronize with the fundamental frequency of the bridge. Under these circumstances large oscillations will be induced and the form in which the bridge then oscillates will approximate very closely to a sine curve of the general type

$$y = A_1 \sin \frac{\pi x}{l}.$$

For short-span bridges, synchronism, even with the fundamental mode of vibration, is out of the question, while on the other hand, for long-span bridges, synchronous oscillations will only be developed at comparatively low speeds, and in such cases a speed limitation may be positively harmful. For medium length bridges of 120 feet span or thereabouts, the fundamental frequency is of the order 6 periods per sec., and resonance will be developed when a locomotive is running more or less at its highest speed and giving hammer-blows of maximum intensity. Accordingly, in the case of railway bridges of this medium-span class, provision must be made for oscillations of exceptional severity.

The analysis of bridge vibrations is considerably complicated by the fact that the hammer-blows producing the oscillations are moving along the bridge, and a most powerful and elegant method of taking this motion into account is provided by the process of harmonic analysis expounded in Chapter I. By this method, any distribution of live load, concentrated or distributed, can, for purposes of calculating deflections, be replaced by an harmonic series of sinusoidal distributions of load, the general form of the series being

$$w_1 \sin \frac{\pi x}{l} + w_2 \sin \frac{2\pi x}{l} + w_3 \sin \frac{3\pi x}{l} + \text{etc. per unit length.}$$

Since, for a beam freely supported at its ends, a sinusoidal distribution of load gives rise to similar sinusoidal distribution of deflection, the simplicity of this process is such that nature obviously intended it to be employed for calculating deflections. In general it will be found that the deflection is almost entirely dominated by the primary component or first term of the harmonic series. Thus for a vertical force P concentrated at a section distant a from an end of the bridge, the deflection is almost exactly that due to a sinusoidal load distribution

$$\frac{2P}{l} \sin \frac{\pi a}{l} \sin \frac{\pi x}{l}$$

per unit length along the span.

For the case $a = l/3$, the degree of accuracy obtained by this method is indicated by Fig. 6, Chapter I. The full line indicates the true deflection and the dotted line is the approximation obtained by replacing the concentrated load by its equivalent first harmonic distribution

$$\frac{2P}{l}\sin\frac{\pi a}{l}\sin\frac{\pi x}{l}.$$

At no point in the span is the discrepancy between these two curves at all serious, and at the centre it can hardly be detected.

If the force is a hammer-blow $P\sin 2\pi Nt$, moving with velocity v along the bridge, then $a=vt$ and the equivalent primary load distribution is

$$\frac{2P}{l}\sin 2\pi Nt\sin\frac{\pi vt}{l}\sin\frac{\pi x}{l}.$$

Putting $n=\dfrac{v}{2l}$, this can be written in the form

$$\frac{P}{l}[\cos 2\pi(N-n)t-\cos 2\pi(N+n)t]\sin\frac{\pi x}{l}.$$

From this emerges the important conclusion that the oscillating effect of a moving force, alternating N times per sec. between the limits $\pm P$, can be viewed as the combined action of two stationary central alternating forces

$$\frac{P}{2}\cos 2\pi(N-n)t \quad \text{and} \quad -\frac{P}{2}\cos 2\pi(N+n)t$$

having slightly different frequencies $N-n$ and $N+n$, and, in the process of these component forces getting into step and out again, the rise and fall of the effectiveness of the moving force as it comes on and leaves the bridge is reproduced with an accuracy which is sufficient for all practical purposes.

If the moving force has a constant magnitude P, the equivalent central operative force is $P\sin 2\pi nt$, where $n=\dfrac{v}{2l}$. For speeds such as are attainable on railway bridges, n is always small compared with the natural frequency of the bridge and, on this account, the effect of a moving force of constant magnitude is almost non-dynamic, the central vertical displacement being almost identical with the "crawl-deflection", that is the central deflection produced when the force is moving quite slowly along the span. The only dynamical effect is a small free oscillation which is developed owing to the fact that, whereas the force comes on to the bridge at a definite speed, the downward deflection must start with zero velocity. In the case of a uniformly distributed smooth-running load even this small state of free oscillation is absent.

The oscillations due to smooth running loads are analysed in Chapter IV, and Chapter V is devoted to the study of the oscillations induced by a moving alternating force $P \sin 2\pi Nt$. Naturally the greatest state of oscillation is set up when resonance is established, that is when N, the frequency of the applied force, coincides with n_0, the fundamental frequency of the bridge.

Fig. 12, Chapter V, is the result of applying analysis to such a case, and this diagram gives the deflection at the centre of the bridge, in terms of the position of the force which is producing the deflection. The bridge in question has a span of about 270 feet and a natural frequency of 3 periods per sec. The load consists of a steady downward force of 100 tons, combined with a periodic force alternating between the limits $\pm 5\cdot4$ tons. The velocity of translation of this combination of forces is about 30 miles per hour, and the frequency of the alternating force being 3 periods per sec., a condition of resonance is established. It will be seen that, under the stimulus of resonance, the oscillations due to the alternating force mount up to an alarming extent, and continue to increase during the whole passage of this force. On the other hand, the effect of the steady force is almost non-dynamic, and the deflection diagram may be viewed as the "crawl-deflection" for the steady load, upon which the state of oscillations due to the alternating force is superposed.

Fortunately this picture is a wild exaggeration of what actually does occur in the case of a railway bridge traversed by a locomotive. It assumes that a perfect condition of resonance can be maintained between the hammer-blows and the fundamental frequency of the bridge during the whole passage of the locomotive, whereas the natural frequency changes continuously with the varying position of the load and, consequently, true resonance can only be momentarily attained. But the real safeguard which shields railway bridges from excessive oscillations is damping, and any theory which leaves this all-important influence out of account will predict oscillations which have little or no connection with reality. This damping influence is made apparent and can be estimated by the rate at which free oscillations die down after a locomotive has passed off a bridge. In long-span bridges the rate of decrease is slow and residual oscillations persist for a considerable time, but in short-span bridges the influence of damping is so pronounced that free oscillations die out almost instantaneously. The method of evaluating this all-important bridge characteristic, and the reason why it is so noticeable in short-span bridges, is dealt with in Chapter XIII and, in that same Chapter, detailed consideration is given to the other bridge and locomotive characteristics which have to be taken into account in estimating bridge oscillations.

For long-span bridges, large oscillations of a synchronous character are associated with comparatively low frequencies, and the vertical accelerations they impart to a locomotive are not sufficient to overcome the friction in its spring movement. Hence for long-span bridges, a locomotive in moving across a bridge behaves as though its springs were locked, and, in estimating its effect in producing bridge oscillations, it is only necessary to know its total mass, the magnitude of its hammer-blow and the size of its driving-wheels.

For the bridge, the characteristics required are the span, the total mass, the unloaded fundamental frequency and the damping coefficient, as deduced from the rate of decrease of residual free oscillations.

The complete analysis for the case of a long-span bridge, in which the oscillations are not sufficiently active to stimulate locomotive spring movement, is set forth in Chapter VIII, and it involves the solution of a somewhat repulsive differential equation, from which arithmetical solutions come not forth save by much prayer and fasting. Applied to particular cases, this analysis is found to yield results which are in excellent agreement with those obtained by practical experiment, and the degree of accuracy achieved is made evident in Figs. 27 (a) and 27 (b).

Fig. 27 (a) is a series of bridge deflection records obtained by the Bridge Stress Committee, and they relate to Newark Dyke Bridge, a single-track bridge having a span of $262\frac{1}{2}$ feet, and an unloaded frequency of 2·88 periods per sec. Its total mass is 460 tons and the locomotive used for this series of tests was a coal locomotive having a mass of 107·65 tons, wheels 14 feet in circumference, and a hammer-blow of 0·576 ton at 1 rev. per sec. When placed at the centre of the span, the mass of the locomotive lowered the natural frequency of the bridge to 2·4 periods per sec. The locomotive was run at speeds ranging from 2·14 to 6·20 revs. per sec. and it will be seen that the greatest oscillations occur when the revolutions are 2·4, that is when the hammer-blows synchronize with the frequency of the bridge at the instant the locomotive is passing the centre of the span. At speeds below the range of loaded frequencies, the oscillations are appreciably reduced, and at very high speeds they become quite insignificant; all of which is in accordance with theoretical predictions.

To this series of tests mathematical analysis was applied, and the results obtained are set forth in Fig. 27 (b). The general agreement is very satisfactory over the whole range of speeds. Theory and experiment agree that the greatest oscillations occur when the revolutions are 2·4, and it will be found that the maximum deflections for this extreme case are almost identical. Experiments, not only in this case, but in others, confirm the conclusion that, for long-span bridges, mathematical analysis can predict the state of oscillation with an accuracy which is quite sufficient for all practical purposes, but the computations are exceedingly laborious, and by no stretch of imagination can the method be regarded as one of general utility. Even at the cost of some slight loss of accuracy, a simplified process must be adopted, and, for the extreme case, when synchronism is established at the instant the locomotive is passing the centre of the span, a comparatively simple formula for predicting the maximum state of oscillation can be evolved, which gives results agreeing almost exactly with those determined by the full and very laborious mathematical analysis.

This method takes full account of the movement of the hammer-blow along the span, but it assumes that the loaded frequency of the bridge remains the same as it is when the locomotive is at the centre of the bridge and that the hammer-blows synchronize with this particular frequency. In consequence of this assumption, the oscillations for the latter part of the passage across the bridge are somewhat exaggerated, but the state of oscillation when the locomotive is passing the centre of the bridge is only overestimated to an insignificant extent.

The assumption of a constant bridge frequency reduces the mathematical analysis to the solution of a straightforward linear differential equation with constant coefficients, and the degree of accuracy achieved is indicated by the two central deflection diagrams which appear in Fig. 28, Chapter VIII.

These diagrams relate to a bridge and locomotive whose characteristics are nearly the same as those previously specified in connection with Figs. 27 (a) and 27 (b), the upper diagram being that given by the full mathematical analysis, and the lower being the result obtained by the approximate method. It will be seen that the agreement is remarkably good, and the over-statement of the maximum increase of the central deflection, as given by the approximate method, is less than 2 per cent.

The formula for determining the semi-amplitude of the maximum central oscillation, as deduced by the approximate process, is

$$\frac{D_P}{2} \cdot \frac{N}{\sqrt{n^2 + n_b'^2}},$$

where D_P is the central deflection due to the hammer-blow at N revs. per sec., applied at the centre of the bridge as a stationary force of constant magnitude, $n = \dfrac{v}{2l}$, N is the natural frequency of the bridge when the locomotive is standing at its centre and n_b' is a constant depending on the damping in the bridge.

From this result it appears that the dynamic effect produced by the movement of the locomotive can be accounted for by an apparent increase in the damping of the bridge.

The proof of this formula and the method of determining n_b' from the residual bridge oscillations is given in the latter part of Chapter VIII.

A comparison of Fig. 12, Chapter V with Fig. 28, Chapter VIII gives ocular evidence of the important part played by damping and the inertia of the locomotive in curbing the exuberance of synchronous oscillations. The upper diagram of Fig. 28 depicts the maximum state of oscillation analysis predicts when these effects are taken into account, and it will be seen that the oscillations are much less vigorous than those shewn in Fig. 12, which gives the corresponding maximum state of oscillation theoretically obtainable when damping and locomotive inertia are idealized out of existence.

It was mentioned earlier on that, even in the absence of hammer-blows, some small amount of oscillation is set up by a moving load owing to its more or less sudden application. How insignificant this is, and how rapidly it dies out owing to damping, is shewn by the upper diagram of Fig. 14, Chapter VI, the only difference between this and that illustrated by the lower diagram being the total absence of hammer-blows. So small are the oscillations that one is almost justified in saying that, for long-span bridges, if hammer-blows are absent, the necessity for impact allowances ceases to exist.

Though simple in composition the approximate formula given above contains all the ingredients it is necessary to include in predicting maximum oscillations in long-span bridges and, broadly speaking, this classification includes bridges of 250 feet span and upwards, in which the oscillations induced by hammer-blows are not, under any normal circumstances, sufficiently energetic to stimulate movement in the springs of the driving-wheels of a locomotive.

For shorter span bridges, the comparatively high frequencies associated with the largest oscillations will increase the vertical accelerations in the bridge to such a pitch that the locomotive springs must inevitably be set in motion, and this movement has a profound effect, both on the magnitude of the bridge oscillations and also on the frequency at which the maximum state of oscillation is developed.

The natural frequency of a locomotive on its springs is usually of the order of 3 periods per sec., and if the bridge is oscillating at an appreciably higher frequency the vertical movements of the spring-borne load and the bridge will be in opposite directions. Hence the mass of the locomotive, instead of lowering the natural frequency of the bridge, may actually raise it above the unloaded frequency. Owing to this, synchronous oscillations in medium-span bridges tend to occur at high engine speeds and, since at high engine speeds hammer-blows are large, the dynamic effects in bridges of this class, say from 100 to 150 feet in span, are particularly pronounced.

Fortunately, however, spring movement also introduces a beneficial influence which partially counteracts dynamic excesses in bridges of medium span. As soon as it occurs, a reservoir is opened capable of absorbing abundance of energy, and the friction between the leaves of the laminated springs, augmented by the friction between the axle-boxes and their guides, provides a damping influence far more potent than any damping in the bridge itself.

Predicting the oscillations in a bridge, due to the passage of an unsprung locomotive by the full mathematical analysis, is not a task to be undertaken in a casual frame of mind, but it is child's play in comparison with the corresponding computation when spring movement has to be taken into account.

This analysis is set forth in Chapter IX, and from it arithmetical results can be deduced; but it is a laborious process and, for the complete analysis of a single case, one must be prepared, even with the assistance of a calculating machine, to face a month's work of concentrated effort. Formidable though it may be, this task cannot be completely shirked, since it provides the only satisfactory method of checking the accuracy of simplified processes, and particularly an approximate method for calculating oscillations in the extreme case when synchronism is established at the instant the locomotive is passing the centre of the span. This approximation is based on the fact that owing to the very heavy extra damping introduced by spring movement, the bridge oscillations mount up approximately to their maximum in a very short space of time. Thus from a particular case studied in Chapter IX it appears that, if a locomotive standing at the centre of the span starts to skid its wheels at a speed which sets up synchronous oscillations, then in one second the oscillations will mount up to 99 per cent. of their full value and at the end of the first half second they will only fall short of this limit by about 10 per cent. For a bridge of 100 feet span or more, a locomotive, even when travelling at its highest possible speed, is in the middle third of the span for at least half a second and, consequently, there is time for the oscillations to build up to very nearly the full value they would attain if the locomotive remained permanently hammering away at the centre of the bridge. Hence, if an estimate of the maximum state of oscillation, set up by a given locomotive on a given bridge of medium span, is required, a good approximation can be obtained by treating the locomotive as stationary at the centre of the bridge, skidding its wheels at the particular speed which sets up synchronous oscillations in the bridge. The mathematical analysis for calculating the oscillations produced by a central skidding locomotive taking into account spring movement is not very complicated. It is set forth in Chapter IX, and from it a comparatively simple formula can be deduced for predicting the greatest dynamical deflection consequent on a locomotive moving along the bridge at the speed which induces the maximum state of oscillation. Tested against a corresponding case worked out by the fullest possible system of mathematical analysis, and compared with bridge records experimentally determined, it appears that the approximate formula may possibly overestimate the maximum state of oscillation by about 10 per cent., but this error, being on the side of safety, may almost be regarded as an advantage.

A somewhat closer approximation can be achieved by taking the hammer-blow of the central skidding locomotive to vary in accordance with the formula

$$\frac{P}{2}\left[\cos 2\pi\left(N-n\right)t - \cos 2\pi\left(N+n\right)t\right],$$

instead of maintaining the constant value $P\sin 2\pi Nt$. As mentioned earlier on, this substitution takes into account the motion of translation of the hammer-blow and makes it possible to predict, with a fair degree of accuracy, the rise and fall of the state of oscillation as the locomotive passes along the span. The degree of accuracy achieved by this approximate process, as tested against the same case worked out by the fullest possible mathematical analysis, is shewn by Fig. 32, Chapter IX, and it will be seen that the agreement leaves little to be desired. But, for computing dynamic allowances, the variation of the state of oscillation is of little practical interest; in general it is only the maximum state of oscillation that is required and, for this purpose, the comparatively simple formula which neglects the motion of translation of the locomotive gives a sufficiently accurate prediction.

This formula requires the same bridge data as in the case of long spans, but the locomotive data must be more detailed. Thus (excluding the tender), the spring-borne mass M_s, and M_u the unsprung mass of the locomotive, must be known separately, also n_s, the frequency of the locomotive on its springs, and F, the value of the total frictional force resisting spring movement.

The evolution of this formula is given in Chapter X. It applies only to bridges of medium span in which synchronous oscillations can be developed, and the expression for the maximum dynamical deflection δ_0 takes the comparatively simple form

$$\delta_0 = D_P \cdot \frac{n_0{}^2}{2N_2 n_b}\left[1 - \frac{4F}{\pi P}\cdot\frac{N_2{}^2}{N_2{}^2 - n_s{}^2}\right],$$

where P is the hammer-blow at N_2 revs. per sec.,

   D_P is the central deflection due to a force P statically applied at the centre,
   n_0 is the unloaded frequency of the bridge,
   n_b is a coefficient which defines the bridge damping,
   N_2 is the hammer-blow frequency which causes the maximum state of oscillation.

The critical frequency N_2 is obtained from the fact that $N_2{}^2$ is the larger of the two roots of the quadratic equation

$$N^4 - N^2 n_1{}^2\left(1 + \frac{n_s{}^2}{n_2{}^2}\right) + n_1{}^2 n_s{}^2 = 0,$$

and, in this equation, if the action of only one locomotive has to be taken into account,

$$n_1{}^2 = \frac{M_G}{M_G + 2M_U}n_0{}^2 \quad \text{and} \quad n_2{}^2 = \frac{M_G}{M_G + 2M_U + 2M_S}n_0{}^2.$$

The principle of replacing a moving locomotive by one skidding its wheels at the centre of the span affords an easy and sufficiently accurate method of predicting oscillations in bridges of medium span, even in cases where synchronous speeds cannot be fully attained.

The application of this principle is explained in Chapter X and experimental justification is provided by Fig. 31, which shews a series of amplitude-frequency observations obtained by the Bridge Stress Committee relating to a medium length bridge of 112 feet span. The observations, though somewhat erratic, can be seen to range themselves along the amplitude-frequency curve deduced by the skidding locomotive method of analysis. The amplitude first of all increases with the frequency until it reaches a value of about 3·5; spring friction then breaks down and, after a slight dip in the curve, the amplitude mounts up again to a maximum when the frequency is a little over 6·5, though, owing to practical limitations of speed, this high-frequency condition of resonance could not be experimentally attained.

For the determination of the maximum state of oscillation in bridges of medium span it is necessary to assign arithmetical values to the two loco-motive characteristics n_s and F. The frequency n_s of a locomotive on its springs can be deduced from a knowledge of the strength of the springs and the mass of the spring-borne part of the locomotive. The friction F in the spring movement is, however, a locomotive characteristic which is much more difficult to evaluate. For a stationary locomotive, it can be deter-mined by plotting the load-displacement curve as the locomotive is loaded up and then unloaded, the vertical intercept between these two curves giving the total frictional force in the spring movement; but, for a locomotive in motion, this frictional resistance is to some extent modified by the altera-tion in the pressure between the axle-boxes and their guides, brought about by the action of inertia and steam pressure.

In the case of the locomotives used in the tests by the Bridge Stress Com-mittee, it appeared that the values of the total frictional force resisting spring movement ranged from 8 to 10 tons; but there is ample room for further research in this direction and, in so much as this particular loco-motive characteristic plays a prominent part in limiting the oscillations in bridges of medium span, finality in the problem of dynamic allowances in railway bridges cannot be achieved, until the nature and value of this somewhat elusive characteristic has been more thoroughly explored.

For long-span bridges an exact knowledge of spring friction is fortunately not required, assuming that it is sufficient to hold the springs locked; for short-span bridges, it is likewise not required, but for quite a different reason. The outstanding characteristic of short-span bridges is the heavy damping their oscillations display. The damping resistances in a bridge are mainly due to lack of elasticity in the track, and to the fact that the track, being continuous, opposes changes of slope at the ends of the bridge. The terminal damping couples thus induced are independent of the length of the bridge, and while their effect on long-span bridges may be slight, their influence on short and comparatively light spans is so pronounced that free oscillations are destroyed with almost instantaneous rapidity.

As shewn in Chapter XIII, for bridges of similar character and with the same type of track, the damping coefficient n_b which figures in the analysis of bridge oscillations varies with the length of the span in a manner given by the relation $n_b = al + \dfrac{b}{l^2}$. The second term of this expression is insignificant for long spans, but for short spans it predominates and causes the value of n_b to rise abruptly.

For short spans of 50 feet or less, the examination of actual records shews that the magnitude of the dynamical deflections can be predicted with sufficient accuracy by treating the hammer-blows as purely static in their effect, but, although no dynamic magnification is apparent, the phase of the oscillations is found to lag behind the phase of the hammer-blows to a considerable extent. Both these circumstances are accounted for by the large amount of damping inherent in short spans. For reasons which are considered in Chapter XII, it causes the operative-forces, that is the fluctuations in the reactions between the wheels and the rails, to become almost equal in magnitude to, though considerably out of phase with, the corresponding hammer-blows, while the heavy bridge damping, combined with the high natural frequency of short spans, makes these operative-forces almost absolutely non-dynamical in their effect. Owing to this combination of circumstances, the determination of the dynamical deflection in short-span bridges, due to hammer-blow, is reduced to a simple statical computation, examples of which are given in Chapter XII. The phase of the oscillations thus deduced, in relation to the phase of the hammer-blows, will be in error, but the magnitude of the oscillations will be correct and, for determining dynamic allowances, this is all that matters.

The degree of accuracy achieved by this simple process is illustrated by Fig. 51, which is a bridge deflection record reproduced from the Bridge Stress Report. It relates to a bridge of about 33 feet span and, as invariably happens in the case of short spans, the smoothness of the record is marred by irregularities, partly due to track imperfections and partly to small higher harmonic components in the operative-forces. The dotted line shews the deflection diagram obtained by treating the hammer-blows as static forces. The maximum deflection, thus deduced, agrees almost exactly with that actually recorded, and the phase difference between the oscillations and the hammer-blows, brought about by bridge damping, is clearly indicated.

When the maximum state of oscillation to which a bridge can be subjected has been determined, the real problem of impact allowances has been solved and the determination of the dynamic forces, bending-moments and shearing-forces consequential on a given state of oscillation is a comparatively straightforward task which might reasonably be left to the bridge designer to include in his calculations. In this connection there is, however, one pitfall to be avoided, which calls for a danger warning. It has been assumed that the mode of oscillation of a bridge is a simple sine curve and, if this assumption was strictly correct, the corresponding distributions of bending-moment and shearing-force would be respectively sine and cosine curves, as deduced by the relationships

$$M = -EI\frac{d^2y}{dx^2} \quad \text{and} \quad S = -EI\frac{d^3y}{dx^3}.$$

Unfortunately this simple process of deduction cannot be employed, because the mode of vibration is not absolutely a pure sine curve and the error contained in this assumption, small as it is, will be magnified into a prominence, which cannot be ignored by the processes of double and treble differentiation. On the other hand, the assumption that the mode of vibration is a simple sine curve introduces no appreciable error in the calculations of the vertical deflections of the bridge and the accelerations and dynamic forces consequent thereon. It is only by referring back to the dynamic forces calculated directly from the bridge deflections that accurate estimates of the dynamic bending-moments and shearing-forces can be determined.

In making these calculations, the systems of forces which have to be taken into account are as follows:

(1) The distribution of force due to the inertia of the bridge.

(2) The distribution of force representing the damping resistances in the bridge.

(3) The concentrated operative-forces, that is, the fluctuations in the reactions between the wheels and the rails.

By combining these various effects, paying due regard to the phase differences which exist, expressions can be formulated specifying the necessary dynamic bending-moment and shearing-force allowances expressed in terms of hammer-blow and the semi-amplitude of the greatest possible oscillation. These expressions are to be found in Chapter XIV.

In calculating impact allowances, only the effect of locomotives need be considered, and the dynamic influence of the train which follows can safely be ignored, since it merely tends to damp down the oscillations set up by hammer-blows.

The tender, passenger coaches, or goods wagons constituting a train, consist of loads which are mainly spring borne, but, unlike a locomotive, they move with comparative freedom on their springs. As the train comes on to a bridge, this spring movement prevents the additional load from having any appreciable effect in altering the natural frequency of the bridge, but it provides some small additional damping which has a beneficial influence.

If the impact allowances imposed take no heed of this advantage, they will only err to a very small extent and the error, such as it is, will have the merit of being on the side of safety.

For a single-track bridge, if the span is sufficiently long, the impact allowances must be worked out on the assumption of two locomotives having their hammer-blows in unison.

For a double-track bridge, the case of a double-heading on each track has to be considered, but the probability of the two pairs of locomotives passing one another at the centre of the span at the same speed and all four hammer-blows being exactly in step at this instant is so infinitely remote, that full allowance need hardly be made for this most improbable combination of circumstances.

In the case of a double-track bridge, if the hammer-blows of the loco-motives on the two tracks are anti-phased, the sides of the bridge will oscillate in opposite directions. This point is discussed in Chapter XIV and the conclusion is reached that the dynamic deflections thus induced in the main girders are likely to be less than those developed when the hammer-blows on the two tracks are in step. Consequently, in formulating impact allowances, there is in general no practical necessity for exploring the in-teresting but somewhat intricate problem of torsional oscillations, though, for calculating the secondary stresses in cross-beams, some consideration must be given to this mode of oscillation. Apart from this, the determination of dynamic stresses in the cross-beams and longitudinal members which con-stitute the deck system of a bridge presents no difficulties. These members usually have such a high natural frequency that the operative-forces they have to support due to hammer-blow may be regarded as non-dynamic in their effect, and impact allowances in such cases can be dealt with by the simple process of superposing these operative forces on the statical loads to which these members are subjected.

In the general mathematical analysis, kinetic units are employed, but, in arithmetical calculations and practical formulae, forces and masses have been expressed in tons.

In addition to the Report of the Bridge Stress Committee previously men-tioned, numerous reports have been published in recent years by the Indian Railway Bridge Committee which give valuable information relating to the experimental determination of impact allowances, but so far as the mathe-matical treatment of this problem is concerned, the field of exploration would appear to have been almost entirely neglected since 1849 when Sir G. G. Stokes published in the *Proceedings* of the Cambridge Philosophical Society his historic paper dealing with the deflection in a girder produced by a concentrated mass moving along the span at a constant speed. Reference, however, should be made to a short article by Professor S. P. Timoshenko published in the *Philosophical Magazine*, Vol. XLIII (1922). By a somewhat different method he arrives at the same result reached in Chapter V, and the present writer, who was then at an early stage in his mathematical investigations, takes this opportunity of recording the stimulus he received from Professor Timoshenko's brief but illuminating contribution.

In conclusion the author wishes to express his thanks to Mr J. A. Gilmore, a research student at the Cambridge Engineering Department, who devoted much time to checking some of the laborious arithmetical computations which appear in this work.

C. E. INGLIS

CAMBRIDGE, *April* 1934

CHAPTER I

DEFLECTIONS OF BEAMS OBTAINED BY METHODS OF HARMONIC ANALYSIS

Consider the case of a beam resting freely on end supports in the manner illustrated by Fig. 1.

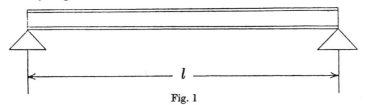

Fig. 1

Suppose that the beam carries a statical distribution of load (including its own weight) which varies along its length, the intensity being w per unit length at a section distant x from the left support. At this same section let M denote the "sagging" bending-moment, S the shearing-force, y the vertical deflection and R the radius of curvature produced by this load

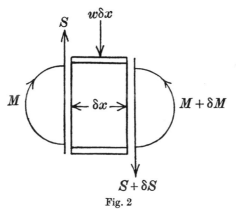

Fig. 2

distribution. By considering the equilibrium of the portion of the beam lying between the sections x and $x + \delta x$, as shewn in Fig. 2,

$$\delta S + w \delta x = 0, \quad \text{i.e.} \quad w = -\frac{dS}{dx},$$

$$\delta M - S \delta x = 0, \quad \text{i.e.} \quad S = \frac{dM}{dx}.$$

Hence
$$w = -\frac{d^2 M}{dx^2}.$$

Moreover $$M = \frac{EI}{R} = -EI\frac{d^2y}{dx^2} \text{ approximately,}$$

where I is the moment of inertia of the cross-section of the beam with reference to its neutral axis and E is Young's Modulus of Elasticity.

Accordingly, $EI\dfrac{d^4y}{dx^4} = w$ is the differential equation for determining the curve of deflection of the beam due to bending-moment.

In what follows it will be assumed that I and E are constants and that the beam is freely supported at its ends, so that y and $\dfrac{d^2y}{dx^2}$ are both zero when $x = 0$ and also when $x = l$.

Suppose the load distribution has the sinusoidal formation specified by

$$w = w_1 \sin\frac{\pi x}{l}.$$

The curve of deflection must satisfy the differential equation

$$EI\frac{d^4y}{dx^4} = w_1 \sin\frac{\pi x}{l}.$$

The solution of this which satisfies the terminal conditions stated above is

$$y = \frac{w_1 l^4}{\pi^4 EI} \sin\frac{\pi x}{l},$$

and the curve of deflection is similar to the load-distribution curve.

If the load distribution is given by $w = w_2 \sin\dfrac{2\pi x}{l}$, the curve of deflection is given by

$$y = \frac{w_2 l^4}{2^4 \pi^4 EI} \sin\frac{2\pi x}{l},$$

and, generally, if the load distribution is $w = w_n \sin\dfrac{n\pi x}{l}$, where n is any whole number, the curve of deflection is a similar sine curve

$$y = \frac{w_n l^4}{n^4 \pi^4 EI} \sin\frac{n\pi x}{l}.$$

Now any load distribution, no matter how irregular and discontinuous it may be, can readily be expressed in the form of an harmonic series of the type

$$w = w_1 \sin\frac{\pi x}{l} + w_2 \sin\frac{2\pi x}{l} + w_3 \sin\frac{3\pi x}{l} + \ldots + w_n \sin\frac{n\pi x}{l} + \ldots.$$

Consequently, on the assumption that the beam is of uniform section and is freely supported at its ends, the curve of deflection is given by the series

$$y = \frac{l^4}{\pi^4 EI}\left[w_1 \sin\frac{\pi x}{l} + \frac{1}{2^4} w_2 \sin\frac{2\pi x}{l} + \frac{1}{3^4} w_3 \sin\frac{3\pi x}{l} + \ldots + \frac{1}{n^4} w_n \sin\frac{n\pi x}{l} + \ldots \right].$$

As the wave-lengths of the component load distributions shorten, their contributions to the total deflection decrease rapidly in importance. This is only natural, since, with a short wave-length, the beam is being subjected to equal positive and negative loads alternating at close intervals along its length.

The harmonious alliance between cause and effect when the load distribution has a sinusoidal formation suggests that, for determining the curve of deflection in a beam of uniform section freely supported at its ends, harmonic analysis is the most incisive and natural process to employ. When once the load distribution has been expressed in the form of an harmonic series, the corresponding curve of deflection can be written down at sight in the form of another harmonic series, and this series converges so rapidly that, for purpose of numerical calculation, only the first two or three components have any real practical importance.

Since this harmonic analysis method for determining deflection permeates all the analytical theory which follows, to make the exposition self-contained, a brief account of the method whereby a given load distribution can be expressed in the form of an harmonic series will now be given.

Suppose a load distribution w per unit length extends over the range $x = 0$ to $x = l$, and that over this range it is required to express w in the form

$$w = w_1 \sin \frac{\pi x}{l} + w_2 \sin \frac{2\pi x}{l} + w_3 \sin \frac{3\pi x}{l} + \ldots + w_n \sin \frac{n\pi x}{l} + \ldots,$$

where n is any whole number.

To obtain w_n multiply both sides of this equation by $\sin \frac{n\pi x}{l}$ and integrate the product between the limits $x = 0$ to $x = l$.

Now $\displaystyle\int_0^l \sin \frac{p\pi x}{l} \sin \frac{n\pi x}{l}\, dx = \frac{1}{2} \int_0^l \left[\cos \frac{(n-p)\pi x}{l} - \cos \frac{(n+p)\pi x}{l} \right] dx$

$$= \frac{l}{2\pi} \left[\frac{1}{n-p} \sin \frac{(n-p)\pi x}{l} - \frac{1}{n+p} \sin \frac{(n+p)\pi x}{l} \right]_0^l,$$

and, if p is a whole number, this is zero except in the special case when $p = n$. For this special case the integral is

$$\int_0^l \sin^2 \frac{n\pi x}{l}\, dx = \frac{1}{2} \int_0^l \left[1 - \cos \frac{2n\pi x}{l} \right] dx$$

$$= \frac{1}{2} \left[x - \frac{l}{2n\pi} \sin \frac{2n\pi x}{l} \right]_0^l$$

$$= \frac{l}{2}.$$

Hence

$$\int_0^l \left[w_1 \sin \frac{\pi x}{l} + w_2 \sin \frac{2\pi x}{l} + \ldots + w_n \sin \frac{n\pi x}{l} + \ldots \right] \sin \frac{n\pi x}{l}\, dx = w_n \frac{l}{2},$$

and consequently

$$w_n = \frac{2}{l} \int_0^l w \sin \frac{n\pi x}{l}\, dx.$$

Evaluating the integrals $\int_0^l w \sin \frac{\pi x}{l}\, dx$, $\int_0^l w \sin \frac{2\pi x}{l}\, dx$, etc., the values of the coefficients w_1, w_2, ... etc. are determined.

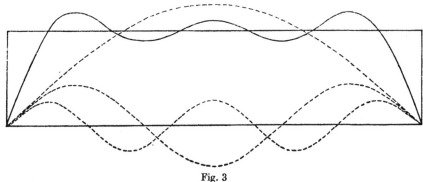

Fig. 3

Case (1)

Suppose w is a constant.

Let $\quad w = w_1 \sin \frac{\pi x}{l} + w_2 \sin \frac{2\pi x}{l} + w_3 \sin \frac{3\pi x}{l} + \ldots + w_n \sin \frac{n\pi x}{l} + \ldots$

between the limits $x = 0$ to $x = l$. Then

$$w_n = \frac{2}{l} \int_0^l w \sin \frac{n\pi x}{l}\, dx = \frac{2w}{l} \times \frac{l}{n\pi} \left[-\cos \frac{n\pi x}{l} \right]_0^l$$

$$= \frac{2w}{n\pi} [1 - \cos n\pi].$$

Hence $w_1 = \frac{4w}{\pi}$, $w_2 = 0$, $w_3 = \frac{4w}{3\pi}$, $w_4 = 0$, $w_5 = \frac{4w}{5\pi}$, etc., and between the limits $x = 0$ to $x = l$, w can be represented by the series

$$\frac{4w}{\pi} \left[\sin \frac{\pi x}{l} + \frac{1}{3} \sin \frac{3\pi x}{l} + \frac{1}{5} \sin \frac{5\pi x}{l} + \ldots \right].$$

Fig. 3 shews the degree of approximation obtained when only the first three terms of the series are taken into account.

Fig. 4 shews the accuracy attained when the series is summed up to 25 terms.

Even when the series is carried this considerable distance the resulting graph presents a corrugated appearance. Its slope in a sense is indeterminate, and this is only natural since differentiation of the harmonic expression for w leads to a non-convergent series which is arithmetically unintelligible.

When a function is expressed by a harmonic series, intelligible conclusions may always be achieved by integrating both sides of the equation, but, in some instances, and the foregoing is a case in point, the series does not admit of differentiation.

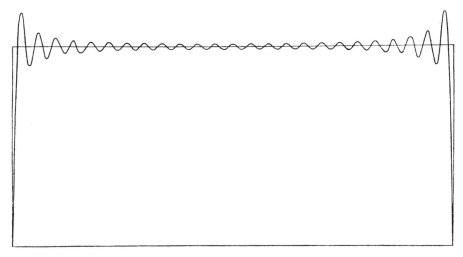

Fig. 4

Case (2)

Suppose the distribution consists of a uniform load per unit length extending over the range limited between $x = a$ to $x = b$, as shewn in Fig. 5.

Let the distribution be expressed by the series

$$w_1 \sin \frac{\pi x}{l} + w_2 \sin \frac{2\pi x}{l} + w_3 \sin \frac{3\pi x}{l} + \ldots + w_n \sin \frac{n\pi x}{l} + \ldots,$$

$$w_n = \frac{2}{l} \int_a^b w \sin \frac{n\pi x}{l} dx = \frac{2w}{n\pi} \Big[-\cos \frac{n\pi x}{l} \Big]_a^b$$

$$= \frac{2w}{n\pi} \Big[\cos \frac{n\pi a}{l} - \cos \frac{n\pi b}{l} \Big] = \frac{4w}{n\pi} \Big[\sin (b+a) \frac{n\pi}{2l} \sin (b-a) \frac{n\pi}{2l} \Big],$$

and the harmonic series for the load distribution takes the form

$$\frac{4w}{\pi} \Big[\sin (b+a) \frac{\pi}{2l} \sin (b-a) \frac{\pi}{2l} \sin \frac{\pi x}{l}$$

$$+ \tfrac{1}{2} \sin 2 (b+a) \frac{\pi}{2l} \sin 2 (b-a) \frac{\pi}{2l} \sin \frac{2\pi x}{l} + \ldots \Big].$$

The important case of a concentrated load W at the section $x = a$ can be derived from this by making $b - a$ very small and putting $w(b-a) = W$. The general term of the above series is

$$\frac{4w}{n\pi} \sin n(b+a)\frac{\pi}{2l} \sin n(b-a)\frac{\pi}{2l} \sin\frac{n\pi x}{l},$$

and, in the limit, when $b = a$ and $w(b-a)$ becomes W, this reduces to

$$\frac{4w}{n\pi} \sin\frac{n\pi a}{l} \times \frac{n(b-a)\pi}{2l} \sin\frac{n\pi x}{l} = \frac{2W}{l} \sin\frac{n\pi a}{l} \sin\frac{n\pi x}{l}.$$

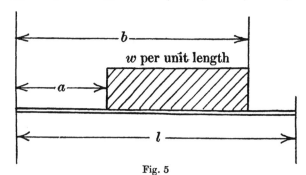

Fig. 5

Hence the harmonic expression for a load W concentrated at the section $x = a$ is

$$\frac{2W}{l}\left[\sin\frac{\pi a}{l} \sin\frac{\pi x}{l} + \sin\frac{2\pi a}{l} \sin\frac{2\pi x}{l} + \sin\frac{3\pi a}{l} \sin\frac{3\pi x}{l} + \dots\right].$$

For a beam of uniform section and of length l, freely supported at its extremities and carrying a load w per unit length uniformly distributed over its whole length, since this load can be expressed in the form

$$\frac{4w}{\pi}\left[\sin\frac{\pi x}{l} + \frac{1}{3}\sin\frac{3\pi x}{l} + \frac{1}{5}\sin\frac{5\pi x}{l} + \dots\right],$$

the corresponding curve of deflection is

$$y = \frac{4wl^4}{\pi^5 EI}\left[\sin\frac{\pi x}{l} + \frac{1}{3^5}\sin\frac{3\pi x}{l} + \frac{1}{5^5}\sin\frac{5\pi x}{l} + \dots\right],$$

and this is practically indistinguishable from

$$y = \frac{4wl^4}{\pi^5 EI} \sin\frac{\pi x}{l},$$

or, in other words, the deflection is almost entirely contributed by the primary component of the load distribution. According to this approximation, the central deflection is

$$\frac{4wl^4}{\pi^5 EI} = \frac{wl^4}{76 \cdot 5 EI}.$$

Actually the central deflection due to bending is

$$\frac{5wl^4}{384EI} = \frac{wl^4}{76\cdot8EI},$$

so the error due to the approximation is less than 0·4 per cent.

For a load W concentrated at the section $x = a$, since the load distribution can be expressed in the form

$$\frac{2W}{l}\left[\sin\frac{\pi a}{l}\sin\frac{\pi x}{l} + \sin\frac{2\pi a}{l}\sin\frac{2\pi x}{l} + \sin\frac{3\pi a}{l}\sin\frac{3\pi x}{l} + \ldots\right],$$

the curve of deflection is

$$y = \frac{2Wl^3}{\pi^4EI}\left[\sin\frac{\pi a}{l}\sin\frac{\pi x}{l} + \frac{1}{2^4}\sin\frac{2\pi a}{l}\sin\frac{2\pi x}{l} + \frac{1}{3^4}\sin\frac{3\pi a}{l}\sin\frac{3\pi x}{l} + \ldots\right].$$

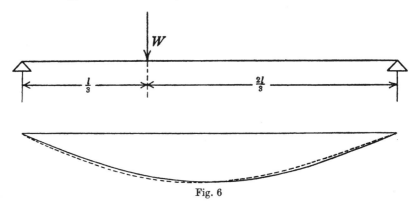

Fig. 6

If the load is near the centre and consequently $\frac{a}{l}$ is nearly $\frac{1}{2}$, the contribution to the total deflection made by the second harmonic component is very small and a close approximation to the curve of deflection is given by

$$y = \frac{2Wl^3}{\pi^4EI}\sin\frac{\pi a}{l}\sin\frac{\pi x}{l}.$$

According to this approximation the central deflection due to a central load W is $\frac{2Wl^3}{\pi^4EI} = \frac{Wl^3}{48\cdot7EI}$, whereas the true deflection due to bending is $\frac{Wl^3}{48EI}$.

Fig. 6 shews the application of this method to the case where $a = \frac{l}{3}$. The dotted line indicates the true deflection and the full line which follows it closely is the result of plotting the first harmonic component of the deflection

$$y = \frac{2Wl^3}{\pi^4EI}\sin\frac{\pi a}{l}\sin\frac{\pi x}{l}.$$

At the centre there is no visible error, and the small discrepancy which appears elsewhere and is indicated by the intercept between the curves can very nearly be accounted for by the second harmonic component

$$y = \frac{2Wl^3}{2^4\pi^4EI} \sin\frac{2\pi a}{l} \sin\frac{2\pi x}{l}.$$

For two equal loads W concentrated at sections at equal distances a from the two ends of the beam the curve of deflection is

$$y = \frac{4Wl^3}{\pi^4EI}\left[\sin\frac{\pi a}{l}\sin\frac{\pi x}{l} + \frac{1}{3^4}\sin\frac{3\pi a}{l}\sin\frac{3\pi x}{l} + \frac{1}{5^4}\sin\frac{5\pi a}{l}\sin\frac{5\pi x}{l} + \dots\right],$$

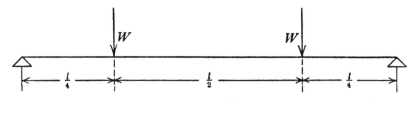

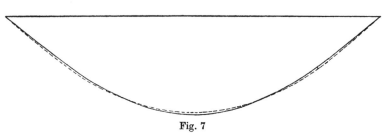

Fig. 7

and consequently even when the loads are a considerable distance from the centre a very close approximation to the curve of deflection is given by

$$y = \frac{4Wl^3}{\pi^4EI}\sin\frac{\pi a}{l}\sin\frac{\pi x}{l}.$$

For the case $a = \frac{l}{4}$, Fig. 7 shews how very closely the true deflection is represented by its first harmonic component, the discrepancy being almost entirely accounted for by the relatively insignificant third harmonic component

$$y = \frac{4Wl^3}{3^4\pi^4EI}\sin\frac{3\pi a}{l}\sin\frac{3\pi x}{l}.$$

Perhaps some critics might be inclined to question the legitimacy of this harmonic process for determining deflections in the case of a concentrated load, owing to the fact that the expansion employed for representing this load is not a convergent series. The most convincing answer to this possible criticism is that the true expression for the curve of deflection produced by a load W at the section $x = a$ is

$$y = \frac{W}{6EI}\left[\{x-a\}^3 + \frac{(l-a)\,x\,(2al - x^2 - a^2)}{l}\right],$$

where the term $\{x-a\}^3$ is only taken into account when $x > a$, and if this expression is harmonically analysed between the limits $x = 0$ to $x = l$ it will be found to have the form

$$y = \frac{2Wl^3}{\pi^4 EI}\left[\sin\frac{\pi a}{l}\sin\frac{\pi x}{l} + \frac{1}{2^4}\sin\frac{2\pi a}{l}\sin\frac{2\pi x}{l} + \ldots\right],$$

which is in agreement with the result previously obtained and establishes the legitimacy of the process adopted.

The conclusion to be drawn from the contents of this Chapter is that, for a beam of uniform section freely supported at its ends, the curve of deflection caused by a load situated in the central part of the beam is a close approximation to a simple sine curve and owes its existence almost entirely to the primary component of the applied load. It is only when the load is near the end of the beam that the second and higher harmonic components of the load need be taken into account, and even then their effect is comparatively small.

For loads disposed symmetrically on either side of the centre of the beam, even when these loads are a considerable distance from the centre, the primary component of the curve of deflection is the only component of any practical importance, since the second harmonic component is necessarily non-existent and the third is relatively insignificant.

The foregoing method of computing deflections has been established for the case of a girder of uniform section, but, since the girders or main trusses supporting railway bridges are not constant in this respect, it is necessary to enquire how far this harmonic analysis method for computing deflections is applicable to girders of non-uniform section.

GIRDER OF NON-UNIFORM SECTION

If w, the load distribution per unit length, is expressed by the harmonic series

$$w = w_1\sin\frac{\pi x}{l} + w_2\sin\frac{2\pi x}{l} + w_3\sin\frac{3\pi x}{l} + \ldots,$$

the curve of deflection is given by the differential equation

$$\frac{d^2}{dx^2}\left(EI\frac{d^2y}{dx^2}\right) = w_1\sin\frac{\pi x}{l} + w_2\sin\frac{2\pi x}{l} + w_3\sin\frac{3\pi x}{l} + \ldots,$$

that is,

$$\frac{d^2y}{dx^2} = \frac{l^2}{\pi^2 EI}\left[w_1\sin\frac{\pi x}{l} + \frac{1}{2^2}w_2\sin\frac{2\pi x}{l} + \frac{1}{3^2}w_3\sin\frac{3\pi x}{l} + \ldots\right].$$

Applying the processes of harmonic analysis to express the variation in $\frac{1}{I}$, let

$$\frac{1}{I} = \frac{1}{I_0}\left[1 + a_2\cos\frac{2\pi x}{l} + a_4\cos\frac{4\pi x}{l} + a_6\cos\frac{6\pi x}{l} + \ldots\right],$$

where $\frac{1}{I_0}$ is the mean value of $\frac{1}{I}$.

Hence the deflection is given by the equation

$$\frac{d^2y}{dx^2} = \frac{l^2}{\pi^2 E I_0}\left[w_1\sin\frac{\pi x}{l} + \frac{w_2}{2^2}\sin\frac{2\pi x}{l} + \frac{w_3}{3^2}\sin\frac{3\pi x}{l} + \ldots\right]$$
$$\times\left[1 + a_2\cos\frac{2\pi x}{l} + a_4\cos\frac{4\pi x}{l} + a_6\cos\frac{6\pi x}{l} + \ldots\right].$$

If the series for w and $\frac{1}{I}$ are limited to the three terms and four terms shewn, then up to the third harmonic component the equation takes the form

$$\frac{d^2y}{dx^2} = \frac{l^2}{\pi^2 E I_0}\left[\left\{w_1\left(1 - \frac{a_2}{2}\right) + \frac{w_3}{3^2}\left(\frac{a_2 - a_4}{2}\right)\right\}\sin\frac{\pi x}{l} + \frac{w_2}{2^2}\left(1 - \frac{a_4}{2}\right)\sin\frac{2\pi x}{l}\right.$$
$$\left. + \left\{w_1\left(\frac{a_2 - a_4}{2}\right) + \frac{w_3}{3^2}\left(1 - \frac{a_6}{2}\right)\right\}\sin\frac{3\pi x}{l}\right],$$

and the solution which satisfies the condition that the ends are freely supported is

$$y = \frac{l^4}{\pi^4 E I_0}\left[\left\{w_1\left(1 - \frac{a_2}{2}\right) + \frac{w_3}{3^2}\left(\frac{a_2 - a_4}{2}\right)\right\}\sin\frac{\pi x}{l} + \frac{w_2}{2^4}\left(1 - \frac{a_4}{4}\right)\sin\frac{2\pi x}{l}\right.$$
$$\left. + \frac{1}{3^2}\left\{w_1\left(\frac{a_2 - a_4}{2}\right) + \frac{w_3}{3^2}\left(1 - \frac{a_6}{2}\right)\right\}\sin\frac{3\pi x}{l}\right].$$

As in the case of a girder of uniform section, the convergency of this series is so rapid that in general an excellent approximation to the curve of deflection is given by the primary harmonic component, which is almost indistinguishable from

$$y = \frac{w_1 l^4}{\pi^4 E I_0}\left(1 - \frac{a_2}{2}\right)\sin\frac{\pi x}{l},$$

and, to this degree of approximation, deflections can be calculated by replacing the non-uniform girder by one of uniform section whose cross-sectional moment of inertia is $I_0\!\left/\!\left(1 - \frac{a_2}{2}\right)\right.$.

By means of a numerical example it will now be shewn that the equivalent moment of inertia $I_0\!\left/\!\left(1 - \frac{a_2}{2}\right)\right.$ differs but little from the moment of inertia of the non-uniform girder, calculated at the centre of the span.

Consider the case in which the moment of inertia of the mid-cross-section is double that at the ends and varies in a parabolic manner between these extreme values.

A close approximation to $\dfrac{1}{I}$ for this case is given by

$$\frac{1}{I}=\frac{1}{I_0}\left[1+0\cdot241\cos\frac{2\pi x}{l}+0\cdot116\cos\frac{4\pi x}{l}+0\cdot066\cos\frac{6\pi x}{l}\right].$$

At the centre $\quad\dfrac{1}{I}=\dfrac{1}{I_0}[1-0\cdot241+0\cdot116-0\cdot066]=\dfrac{0\cdot809}{I_0}$,

and accordingly the moment of inertia for the mid-cross-section is $1\cdot23I_0$.

The moment of inertia for the equivalent girder of uniform section, as given by the expression $I_0\Big/\left(1-\dfrac{a_2}{2}\right)$, for this case, has the value

$$\frac{I_0}{1-0\cdot121}=1\cdot14I_0,$$

which only differs from the maximum moment of inertia by about 8 per cent.

Consequently, if the equivalent girder of uniform section is assumed to have the moment of inertia of the mid-cross-section of a bridge, the deflections so deduced will be under-estimated to a slight extent, but the discrepancy is too small to be of any real practical importance, and, to some extent it compensates for the fact that calculated bridge deflections are apt to be over-estimated owing to the neglect of the additional stiffness contributed by the longitudinal members of the deck system, which is not usually taken into account.

CHAPTER II

NATURAL FREQUENCIES AND NORMAL MODES OF VIBRATION

The extent to which a flexible structure such as a bridge will respond to the action of a pulsating force largely depends on how nearly the frequency of the pulsations coincides with the natural frequency of the structure. Consequently, as a preliminary to examining the state of oscillation forced in a bridge by moving or pulsating forces, a knowledge of its natural frequencies and normal modes of vibration is essential.

Consider a girder of uniform mass and section freely supported at its extremities. Let m be its mass per unit length, and suppose that when set into a state of vertical oscillation α is the vertical downward acceleration at a section distant x from one end. The effect of inertia of the girder is equivalent to an upward distribution of force amounting to $m\alpha$ per unit length and, leaving out of account gravity forces and damping actions, the equation of motion for the beam is

$$EI\frac{d^4y}{dx^4} = -m\alpha,$$

and since $\alpha = \frac{d^2y}{dt^2}$, the equation takes the form

$$EI\frac{d^4y}{dx^4} + m\frac{d^2y}{dt^2} = 0.$$

Write this
$$\frac{d^4y}{dx^4} + K^2\frac{d^2y}{dt^2} = 0, \text{ where } K^2 = \frac{m}{EI}.$$

Consider a solution of the form

$$y = f(t)\sin\frac{\pi x}{l}.$$

This solution satisfies the terminal conditions that y and $\frac{d^2y}{dx^2}$ should both be zero when $x = 0$ and also when $x = l$. It will fit the differential equation if

$$\frac{\pi^4}{l^4}f(t) + K^2\frac{d^2f}{dt^2} = 0.$$

Hence
$$f(t) = A_1\sin 2\pi n_0 t,$$

where
$$2\pi n_0 = \frac{\pi^2}{l^2}\cdot\frac{1}{K} = \frac{\pi^2}{l^2}\sqrt{\frac{EI}{m}}.$$

A possible state of free oscillation is accordingly given by the equation

$$y = A_1 \sin \frac{\pi x}{l} \sin 2\pi n_0 t.$$

This is called the fundamental mode of vibration, and n_0 is termed the fundamental frequency.

Another mode of vibration which also satisfies the terminal conditions can be obtained by assuming

$$y = f(t) \sin \frac{2\pi x}{l},$$

where $f(t)$ must satisfy the equation

$$\frac{2^4 \pi^4}{l^4} f(t) + K^2 \frac{d^2 f}{dt^2} = 0.$$

Hence
$$f(t) = A_2 \sin 2^2 (2\pi n_0 t),$$

and this establishes the existence of another possible mode of vibration of the type

$$y = A_2 \sin \frac{2\pi x}{l} \sin 2^2 (2\pi n_0 t).$$

In this way a series of possible modes of vibration are brought to light of the form

$$y = A_1 \sin \frac{\pi x}{l} \sin 2\pi n_0 t,$$

$$y = A_2 \sin \frac{2\pi x}{l} \sin 2^2 (2\pi n_0 t),$$

$$y = A_3 \sin \frac{3\pi x}{l} \sin 3^2 (2\pi n_0 t),$$

$$\dotsb\dotsb\dotsb\dotsb\dotsb$$

$$y = A_n \sin \frac{n\pi x}{l} \sin n^2 (2\pi n_0 t),$$

where $2\pi n_0 = \frac{\pi^2}{l^2} \sqrt{\frac{EI}{m}}$ defines n_0, the fundamental frequency, the frequencies of the higher modes of vibration being $2^2 n_0$, $3^2 n_0$, etc.

NUMERICAL EXAMPLE

Consider the fundamental frequency of a $24'' \times 7\frac{1}{2}''$ rolled steel joist, 30 feet long, resting freely on end supports. The joist weighs 90 lb. per foot run, and I for its cross-section is 2443 inch⁴. Take $E = 30 \times 10^6$ lb. per square inch.

In working out the formula $2\pi n_0 = \frac{\pi^2}{l^2} \sqrt{\frac{EI}{m}}$, lengths must be expressed in feet and forces in poundals. Hence

$$2\pi n_0 = \frac{\pi^2}{30 \times 30} \sqrt{30 \times 10^6 \times 144 \times g \times \frac{2443}{(144)^2} \frac{1}{90}},$$

i.e.
$$n_0 = \frac{\pi \times 10^3}{12 \times 18 \times 10^2} \sqrt{26221 \cdot 5} = 23 \cdot 56.$$

The second mode of vibration will accordingly have a frequency of 94·24 cycles per sec.

In the previous Chapter it was shewn that a very close approximation to the central deflection D_W produced in a girder by a steady central load W was given by
$$D_W = \frac{2Wl^3}{\pi^4 EI},$$

but
$$\pi^4 EI = 4\pi^2 n_0{}^2 ml^4.$$

Hence
$$D_W = \frac{2Wl^3}{4\pi^2 n_0{}^2 ml^4} = \frac{W}{2\pi^2 n_0{}^2 M_G},$$

where M_G is the total mass of the girder.

By means of this formula the fundamental frequency of a girder can be expressed in terms of the central deflection produced by a given central load and, in the case of railway bridges, this is a most convenient method for obtaining this highly important bridge characteristic.

Railway bridges, of course, are not in general of uniform mass and section, but for purposes of calculating deflections and natural frequencies they can without serious loss of accuracy be subjected to this idealization. If m is taken to be the average mass per unit length of the bridge and I the moment of inertia of the mid-cross-section, the foregoing formulae for determining natural frequencies will yield results which differ but little from the truth. The determination of natural frequencies of girders of varying mass and stiffness admits of exact analytical treatment, but the difference of the results so obtained and those derived on the assumption that the bridge is of uniform mass and section is so slight that it hardly justifies the considerable extra labour involved in the application of the exact process.

A matter of much greater importance is the alteration in the fundamental frequency of a bridge brought about by a mass which it may be carrying and which participates in the vertical oscillations of the bridge.

NATURAL FREQUENCY OF A LOADED GIRDER

Let M be the mass concentrated at a section defined by $x = a$. If α is the downward vertical acceleration of the mass, the girder experiences an upward force $M\alpha$ concentrated at the section $x = a$. The primary harmonic component of this force is

$$\frac{2M\alpha}{l} \sin \frac{\pi a}{l} \sin \frac{\pi x}{l},$$

and, assuming that the other harmonic components of this force make negligible contributions to the deflection, the state of free oscillation is given by the equation

$$EI \frac{d^4 y}{dx^4} = -m \frac{d^2 y}{dt^2} - \frac{2M\alpha}{l} \sin \frac{\pi a}{l} \sin \frac{\pi x}{l},$$

or

$$EI \frac{d^4 y}{dx^4} + m \frac{d^2 y}{dt^2} = -\frac{2M\alpha}{l} \sin \frac{\pi a}{l} \sin \frac{\pi x}{l}.$$

For a solution take

$$y = f(t) \sin \frac{\pi x}{l},$$

then

$$\alpha = \frac{d^2 f}{dt^2} \sin \frac{\pi a}{l},$$

and the equation for $f(t)$ takes the form

$$EI \frac{\pi^4}{l^4} f(t) + \left(m + \frac{2M}{l} \sin^2 \frac{\pi a}{l} \right) \frac{d^2 f}{dt^2} = 0.$$

Hence

$$f(t) = A \sin 2\pi n_1 t,$$

where

$$2\pi n_1 = \frac{\pi^2}{l^2} \sqrt{\frac{EI}{m + \frac{2M}{l} \sin^2 \frac{\pi a}{l}}}$$

$$= \frac{\pi^2}{l^2} \sqrt{\frac{EIl}{M_G + 2M \sin^2 \frac{\pi a}{l}}},$$

where M_G is the total mass of the girder.

The presence of the mass M lowers the fundamental frequency and the reduction corresponds to the addition of a load $2M \sin^2 \frac{\pi a}{l}$ uniformly distributed along the girder.

The reduction depends upon the position of the load, and as a mass such as a locomotive moves along a bridge the natural frequency of the bridge changes continuously. This fact has a most important bearing on the production of bridge oscillations, and this variation of fundamental frequency has the beneficial effect of shielding railway bridges from experiencing the dangers consequent on a complete and continued state of resonance.

Up to now no account has been taken of damping. In railway bridges the oscillations are always damped, as can be seen by the more or less rapid subsidence of the oscillations left in a bridge after the passage of a locomotive. The damping is caused partly by imperfect elasticity in the structure, augmented by the lack of elasticity in the track and ballast. Other sources of damping are to be found in friction at the end bearings and the terminal constraints produced in the bridge by the continuity of the track. The effect is most pronounced in short-span bridges, where the damping is often so heavy that residual oscillations, after the passage of a locomotive, can hardly be detected. In long-span bridges these residual oscillations persist for a considerable length of time, but even in such cases damping must be taken into account for, small though it may be, it exerts a powerful influence in checking the exuberance of synchronous oscillations.

The effect of damping is dealt with by treating it as a resistance to movement distributed along the bridge. The magnitude of the resistance at any section is taken to be proportional to the velocity of movement of that section and the resistance per unit length is accordingly expressed in the form $b\dfrac{dy}{dt}$. For the unloaded bridge the equation giving a state of free oscillation is accordingly

$$EI\frac{d^4y}{dx^4} = -m\frac{d^2y}{dt^2} - b\frac{dy}{dt},$$

or

$$EI\frac{d^4y}{dx^4} + b\frac{dy}{dt} + m\frac{d^2y}{dt^2} = 0.$$

For the fundamental mode of vibration assume $y = f(t)\sin\dfrac{\pi x}{l}$, and the equation for determining $f(t)$ takes the form

$$EI\frac{\pi^4}{l^4}f(t) + b\frac{df}{dt} + m\frac{d^2f}{dt^2} = 0.$$

Now $EI\dfrac{\pi^4}{l^4} = 4\pi^2 n_0^2 m$, where n_0 is the fundamental frequency of the unloaded and undamped bridge.

Let $b = 4\pi n_b m$, where n_b has the dimensions of a frequency. The equation for $f(t)$ then assumes the form

$$\frac{d^2f}{dt^2} + 4\pi n_b\frac{df}{dt} + 4\pi^2 n_0^2 f(t) = 0,$$

and provided that $n_b < n_0$,

$$f(t) = e^{-\alpha t}[A \sin \beta t + B \cos \beta t],$$

where

$$\alpha = 2\pi n_b,$$

$$\beta = 2\pi \sqrt{n_0{}^2 - n_b{}^2},$$

and A and B are independent constants of integration.

In all practical cases relating to bridge oscillations $n_b{}^2$ is negligible in comparison with $n_0{}^2$, and, to a high degree of accuracy, the fundamental mode of vibration can be written

$$y = e^{-2\pi n_b t}[A \sin 2\pi n_0 t + B \cos 2\pi n_0 t] \sin \frac{\pi x}{l}.$$

The natural frequency to this degree of approximation is unaffected by damping, but successive oscillations die down in a Geometric Progression, and, since $\dfrac{1}{n_0}$ is the time of a complete oscillation, the common ratio of this series is $e^{-2\pi \frac{n_b}{n_o}}$. By noting the rapidity with which residual oscillations in a bridge subside, a numerical value can be assigned to n_b.

NUMERICAL EXAMPLE

For a bridge of 250 feet span a reasonable value for the common ratio between successive oscillations is found to be 0·85 and n_0 is of the order 3. Taking these values

$$e^{-2\pi \frac{n_b}{n_o}} = \frac{8 \cdot 5}{10},$$

$$2\pi \frac{n_b}{n_0} \log_{10} e = 1 - \log_e 8 \cdot 5,$$

$$\frac{n_b}{n_0}[2 \cdot 73] = 0 \cdot 07058,$$

$$\frac{n_b}{n_0} = 0 \cdot 0257,$$

$$n_b = 0 \cdot 077, \text{ which is quite small in comparison with } n_0.$$

For a bridge carrying a stationary load it can be shewn that the fundamental damped mode of vibration is given approximately by

$$y = e^{-2\pi n_b t}[A \sin 2\pi n_1 t + B \cos 2\pi n_1 t] \sin \frac{\pi x}{l},$$

where n_1 is the fundamental frequency of the loaded but undamped bridge.

It is hardly necessary to examine the higher modes of vibration. In railway bridges the frequencies of these are generally so far above those of the oscillations forced in the bridge by the passage of a locomotive, that they are never stimulated into existence to any appreciable extent and they can in consequence be dismissed from further consideration.

NATURAL FREQUENCIES AND MODES OF OSCILLATION FOR A GIRDER OF NON-UNIFORM SECTION

The equation for a free undamped oscillation is

$$\frac{d^2}{dx^2}\left[EI\frac{d^2y}{dx^2}\right]+m\frac{d^2y}{dt^2}=0.$$

In railway bridges, though I will vary considerably, the variation in m is much less and, in the following analysis, m will be treated as constant.

Following the procedure adopted in Chapter I, $\frac{1}{I}$ is expressed in the form of the harmonic series

$$\frac{1}{I}=\frac{1}{I_0}\left[1+a_2\cos\frac{2\pi x}{l}+a_4\cos\frac{4\pi x}{l}+a_6\cos\frac{6\pi x}{l}+...\right].$$

Let
$$y=\left[A_1\sin\frac{\pi x}{l}+A_3\sin\frac{3\pi x}{l}+A_5\sin\frac{5\pi x}{l}+...\right]\sin 2\pi Nt,$$

then the differential equation for y takes the form

$$\frac{d^2y}{dx^2}=-\frac{4\pi^2N^2ml^2}{\pi^2EI_0}\left[A_1\sin\frac{\pi x}{l}+\frac{A_3}{3^2}\sin\frac{3\pi x}{l}+\frac{A_5}{5^2}\sin\frac{5\pi x}{l}+...\right]$$
$$\times\left[1+a_2\cos\frac{2\pi x}{l}+a_4\cos\frac{4\pi x}{l}+a_6\cos\frac{6\pi x}{l}+...\right]\sin 2\pi Nt.$$

If the series for y and $\frac{1}{I}$ are limited to three and four terms respectively, then up to the fifth harmonic component the equation takes the form

$$\frac{d^2y}{dx^2}=-\frac{4\pi^2N^2ml^2}{\pi^2EI_0}\left[\left\{A_1\left(1-\frac{a_2}{2}\right)+\frac{A_3}{3^2}\left(\frac{a_2-a_4}{2}\right)+\frac{A_5}{5^2}\left(\frac{a_4-a_6}{2}\right)\right\}\sin\frac{\pi x}{l}\right.$$
$$+\left\{A_1\left(\frac{a_2-a_4}{2}\right)+\frac{A_3}{3^2}\left(1-\frac{a_6}{2}\right)+\frac{A_5}{5^2}\cdot\frac{a_2}{2}\right\}\sin\frac{3\pi x}{l}$$
$$\left.+\left\{A_1\left(\frac{a_4-a_6}{2}\right)+\frac{A_3}{3^2}\cdot\frac{a_2}{2}+\frac{A_5}{5^2}\right\}\sin\frac{5\pi x}{l}\right]\sin 2\pi Nt.$$

Hence

$$y=\frac{4\pi^2N^2ml^4}{\pi^4EI_0}\left[\left\{A_1\left(1-\frac{a_2}{2}\right)+\frac{A_3}{3^2}\left(\frac{a_2-a_4}{2}\right)+\frac{A_5}{5^2}\left(\frac{a_4-a_6}{2}\right)\right\}\sin\frac{\pi x}{l}\right.$$
$$+\frac{1}{3^2}\left\{A_1\left(\frac{a_2-a_4}{2}\right)+\frac{A_3}{3^2}\left(1-\frac{a_6}{2}\right)+\frac{A_5}{5^2}\cdot\frac{a_2}{2}\right\}\sin\frac{3\pi x}{l}$$
$$\left.+\frac{1}{5^2}\left\{A_1\left(\frac{a_4-a_6}{2}\right)+\frac{A_3}{3^2}\cdot\frac{a_2}{2}+\frac{A_5}{5^2}\right\}\sin\frac{5\pi x}{l}\right]\sin 2\pi Nt,$$

and, since $\quad y = \left(A_1 \sin \dfrac{\pi x}{l} + A_3 \sin \dfrac{3\pi x}{l} + A_5 \sin \dfrac{5\pi x}{l} \right) \sin 2\pi Nt,$

by equating the coefficients of $\sin \dfrac{\pi x}{l}$, $\sin \dfrac{3\pi x}{l}$ and $\sin \dfrac{5\pi x}{l}$ on the two sides of the equation, three conditions are obtained from which N^2 and the ratios $A_1 : A_3 : A_5$ can be determined.

NUMERICAL EXAMPLE

Consider the case studied in Chapter I, in which

$$\frac{1}{I} = \frac{1}{I_0} \left[1 + 0 \cdot 241 \cos \frac{2\pi x}{l} + 0 \cdot 116 \cos \frac{4\pi x}{l} + 0 \cdot 066 \cos \frac{6\pi x}{l} \right],$$

that is $\qquad a_2 = 0 \cdot 241, \quad a_4 = 0 \cdot 116, \quad a_6 = 0 \cdot 066.$

Let $\dfrac{4\pi^2 N^2 m l^4}{\pi^4 E I_0}$ be denoted by C, then for this case

$$y = C \left[(0 \cdot 879 A_1 + 0 \cdot 007 A_3 + 0 \cdot 001 A_5) \sin \frac{\pi x}{l} \right.$$

$$+ (0 \cdot 007 A_1 + 0 \cdot 012 A_3 + 0 \cdot 0005 A_5) \sin \frac{3\pi x}{l}$$

$$\left. + (0 \cdot 001 A_1 + 0 \cdot 0005 A_3 + 0 \cdot 0016 A_5) \sin \frac{5\pi x}{l} \right] \sin 2\pi Nt.$$

Equating coefficients of $\sin \dfrac{\pi x}{l}$, $\sin \dfrac{3\pi x}{l}$ and $\sin \dfrac{5\pi x}{l}$ on both sides of this equation, the following conditions are obtained:

$$\left(0 \cdot 879 - \frac{1}{C} \right) A_1 + 0 \cdot 007 A_3 + 0 \cdot 001 A_5 = 0,$$

$$0 \cdot 007 A_1 + \left(0 \cdot 012 - \frac{1}{C} \right) A_3 + 0 \cdot 0005 A_5 = 0,$$

$$0 \cdot 001 A_1 + 0 \cdot 0005 A_3 + \left(0 \cdot 0016 - \frac{1}{C} \right) A_5 = 0.$$

By eliminating A_1, A_3 and A_5 from these equations, three values of C can be found, the least value being given by $C = \dfrac{1}{0 \cdot 861}$, and the corresponding fundamental mode of oscillation is

$$y = A_1 \left[\sin \frac{\pi x}{l} - 0 \cdot 0023 \sin \frac{3\pi x}{l} - 0 \cdot 0003 \sin \frac{5\pi x}{l} \right].$$

It thus appears that the fundamental mode of oscillation is quite indistinguishable from a pure sine curve $y = A \sin \dfrac{\pi x}{l}$. The frequency n_0 of the oscillation is given by

$$2\pi n_0 = \frac{\pi^2}{l^2} \sqrt{\frac{E I_0}{m}} \cdot C,$$

and since $C = \dfrac{1}{0\cdot 861}$ and $I_0 = \dfrac{I}{1\cdot 23}$, where I is the moment of inertia of the mid-cross-section,

$$2\pi n_0 = \frac{\pi^2}{l^2} \sqrt{\frac{E I}{m} \cdot \frac{1}{1\cdot 23 \times 0\cdot 861}} = 1\cdot 02 \frac{\pi^2}{l^2} \sqrt{\frac{E I}{m}}.$$

Hence the simple and satisfactory conclusion is reached that the fundamental frequency and mode of oscillation of a bridge can be obtained with great accuracy by idealizing the bridge as a girder of uniform section, having the same sectional moment of inertia as that of the mid-cross-section of the bridge. This idealization will be adopted throughout all the following analysis.

CHAPTER III

OSCILLATIONS PRODUCED BY STATIONARY BUT ALTERNATING DISTRIBUTIONS OF LOAD

As in the previous Chapters the girder under consideration is assumed to be of uniform mass and section freely supported at its ends.

Suppose it is set into a state of oscillation by a load distributed in a sinusoidal manner which alternates n times per second.

Let $w = w_1 \sin \dfrac{\pi x}{l} \sin 2\pi n t$ define this load.

Neglecting damping and the effect of gravity forces, the differential equation for the motion is

$$EI\frac{d^4y}{dx^4} + m\frac{d^2y}{dt^2} = w_1 \sin \frac{\pi x}{l} \sin 2\pi n t.$$

For the particular integral of this equation take

$$y = A \sin \frac{\pi x}{l} \sin 2\pi n t.$$

This solution will fit the equation if

$$A\left[EI\frac{\pi^4}{l^4} - 4\pi^2 n^2 m \right] = w_1,$$

i.e. if

$$A\left[1 - \frac{4\pi^2 n^2 m}{EI\dfrac{\pi^4}{l^4}} \right] = \frac{w_1 l^4}{\pi^4 EI},$$

i.e. if

$$A\left[1 - \frac{n^2}{n_0^{\,2}} \right] = \frac{w_1 l^4}{\pi^4 EI}.$$

Hence the particular integral, or "forced oscillation" as it will in future be called, is given by

$$y = \frac{w_1 l^4}{\pi^4 EI\left(1 - \dfrac{n^2}{n_0^{\,2}} \right)} \sin \frac{\pi x}{l} \sin 2\pi n t.$$

It has been shewn previously that the statical deflection produced by a steadily applied load $w_1 \sin \dfrac{\pi x}{l}$ is given by

$$y = \frac{w_1 l^4}{\pi^4 EI} \sin \frac{\pi x}{l}.$$

Hence the dynamical magnification due to the alternating character of the load is

$$\frac{1}{1 - \dfrac{n^2}{n_0{}^2}}.$$

The oscillations are in phase with or anti-phased to the load according as n is less or greater than n_0. If $n = n_0$, in the absence of damping, the oscillations tend to mount up to an indefinite extent. This process, however, takes time, and the rapidity with which these oscillations mount up is a point which demands attention.

Suppose that when $t = 0$ the girder is at rest with zero deflection. The complete solution of the equation

$$EI\frac{d^4y}{dx^4} + m\frac{d^2y}{dt^2} = w_1 \sin\frac{\pi x}{l}\sin 2\pi n t$$

consists of the forced oscillation which has just been obtained combined with a so-called complementary function which is a solution of the equation

$$EI\frac{d^4y}{dx^4} + m\frac{d^2y}{dt^2} = 0.$$

This is the equation for a free oscillation of the girder, and in future the complementary function will be referred to as a free oscillation. It has been shewn that

$$y = A\sin\frac{\pi x}{l}\sin 2\pi n_0 t$$

is a possible free oscillation, where

$$2\pi n_0 = \frac{\pi^2}{l^2}\sqrt{\frac{EI}{m}}.$$

Superposing a free oscillation of this type on the forced oscillation previously established, a solution which fits the differential equation and at the same time satisfies the starting conditions, viz. $y = 0$ and $\dfrac{dy}{dt} = 0$ when $t = 0$, can be obtained in the form

$$y = \frac{w_1 l^4}{\pi^4 EI\left(1 - \dfrac{n^2}{n_0{}^2}\right)}\sin\frac{\pi x}{l}\left[\sin 2\pi n t - \frac{n}{n_0}\sin 2\pi n_0 t\right].$$

The motion accordingly consists of the superposition of two oscillations, one of which has the frequency of the applied force and the other the fundamental natural frequency of the girder.

The particular case of resonance when $n = n_0$ calls for special treatment, because in this case the expression for y assumes an indeterminate form $\frac{0}{0}$.

To evaluate this form, put $n = n_0 + \delta n$, where, in the limit, δn is made zero. The expression for y then takes the form

$$y = \frac{w_1 l^4 \sin \frac{\pi x}{l}}{\pi^4 EI \left[1 - \left(1 + \frac{\delta n}{n_0} \right)^2 \right]} \left[\sin 2\pi (n_0 + \delta n) t - \left(1 + \frac{\delta n}{n_0} \right) \sin 2\pi n_0 t \right].$$

Neglecting powers of $\frac{\delta n}{n_0}$ above the first,

$$y = \frac{w_1 l^4 \sin \frac{\pi x}{l}}{\pi^4 EI \left[-\frac{2\delta n}{n_0} \right]} \left[\sin 2\pi n_0 t + 2\pi \delta n t \cos 2\pi n_0 t - \sin 2\pi n_0 t - \frac{\delta n}{n_0} \sin 2\pi \delta n_0 t \right],$$

i.e.
$$y = \frac{w_1 l^4 \sin \frac{\pi x}{l}}{\pi^4 EI} \left[\tfrac{1}{2} \sin 2\pi n_0 t - \pi n_0 t \cos 2\pi n_0 t \right].$$

The dynamic magnification after a time t approximates to $\pi n_0 t$, and when $\pi n_0 t$ is large compared with $\tfrac{1}{2}$ the oscillations mount up at a uniform time rate and are in quadrature with the phase of the applied load.

If the distributed alternating load is

$$w = w_2 \sin \frac{2\pi x}{l} \sin 2\pi n t,$$

the forced oscillation is

$$y = \frac{w_2 l^4}{\pi^4 EI \left[2^4 - \frac{n^2}{n_0{}^2} \right]} \sin \frac{2\pi x}{l} \sin 2\pi n t,$$

and the free oscillation of this same mode is of the form

$$y = A \sin \frac{2\pi x}{l} \sin 2^2 (2\pi n_0 t).$$

A superposition of these to satisfy the starting conditions leads to the state of oscillation

$$y = \frac{w_2 l^4}{\pi^4 EI \left[2^4 - \frac{n^2}{n_0{}^2} \right]} \sin \frac{2\pi x}{l} \left[\sin 2\pi n t - \frac{n}{2^2 n_0} \sin 2^2 (2\pi n_0 t) \right].$$

For the general case where the alternating load is expressed by the harmonic series

$$w = \left[w_1 \sin\frac{\pi x}{l} + w_2 \sin\frac{2\pi x}{l} + w_3 \sin\frac{3\pi x}{l} + \dots \right] \sin 2\pi n t,$$

the state of oscillation built up from a condition of rest and zero deflection is given by

$$y = \frac{l^4}{\pi^4 EI} \left[\frac{w_1 \sin\dfrac{\pi x}{l}}{1 - \dfrac{n^2}{n_0{}^2}} \left\{ \sin 2\pi n t - \frac{n}{n_0} \sin 2\pi n_0 t \right\} \right.$$

$$+ \frac{w_2 \sin\dfrac{2\pi x}{l}}{2^4 - \dfrac{n^2}{n_0{}^2}} \left\{ \sin 2\pi n t - \frac{n}{2^2 n_0} \sin 2^2 (2\pi n_0 t) \right\}$$

$$\left. + \frac{w_3 \sin\dfrac{3\pi x}{l}}{3^4 - \dfrac{n^2}{n_0{}^2}} \left\{ \sin 2\pi n t - \frac{n}{3^2 n_0} \sin 3^2 (2\pi n_0 t) \right\} + \dots \right].$$

Particular cases

For a uniformly distributed load w per unit length extending over the whole length of the girder, it has been shewn that

$$w_1 = \frac{4w}{\pi}, \quad w_2 = 0, \quad w_3 = \frac{4w}{3\pi}, \quad w_4 = 0, \quad \text{etc.}$$

Hence the state of oscillation built up in time t by this uniformly distributed load alternating n times per second is

$$y = \frac{4wl^4}{\pi^5 EI} \left[\frac{\sin\dfrac{\pi x}{l}}{1 - \dfrac{n^2}{n_0{}^2}} \left\{ \sin 2\pi n t - \frac{n}{n_0} \sin 2\pi n_0 t \right\} \right.$$

$$\left. + \frac{1}{3} \cdot \frac{\sin\dfrac{3\pi x}{l}}{3^4 - \dfrac{n^2}{n_0{}^2}} \left\{ \sin 2\pi n t - \frac{n}{3^2 n_0} \sin 3^2 (2\pi n_0 t) \right\} + \dots \right],$$

and, generally, a sufficiently good approximation is given by the first harmonic component, that is, by

$$y = \frac{4wl^4}{\pi^5 EI} \left[\frac{\sin\dfrac{\pi x}{l}}{1 - \dfrac{n^2}{n_0{}^2}} \left\{ \sin 2\pi n t - \frac{n}{n_0} \sin 2\pi n_0 t \right\} \right];$$

if n is small compared with n_0, the oscillation approximates to

$$y = \frac{4wl^4}{\pi^5 EI}\left[\sin\frac{\pi x}{l} + \frac{1}{3^5}\sin\frac{3\pi x}{l} + \frac{1}{5^5}\sin\frac{5\pi x}{l} + \ldots\right]\sin 2\pi nt,$$

and this is merely the deflection which would be produced if the load was treated as a statical distribution of force. If n is large in comparison with n_0, the primary component of the oscillation sinks into insignificance, but one of the higher harmonic components may be stimulated into activity.

For a load W concentrated at the section $x = a$, it has been established that

$$w_1 = \frac{2W}{l}\sin\frac{\pi a}{l}, \quad w_2 = \frac{2W}{l}\sin\frac{2\pi a}{l}, \quad \text{etc.}$$

Consequently, the state of oscillation built up in time t by this force when alternating n times per second is given by

$$y = \frac{2Wl^3}{\pi^4 EI}\left[\frac{\sin\dfrac{\pi a}{l}\sin\dfrac{\pi x}{l}}{1 - \dfrac{n^2}{n_0^2}}\left\{\sin 2\pi nt - \frac{n}{n_0}\sin 2\pi n_0 t\right\}\right.$$
$$\left. + \frac{\sin\dfrac{2\pi a}{l}\sin\dfrac{2\pi x}{l}}{2^4 - \dfrac{n^2}{n_0^2}}\left\{\sin 2\pi nt - \frac{n}{2^2 n_0}\sin 2^2\left(2\pi n_0 t\right)\right\} + \ldots\right].$$

In this Chapter the influence of damping has purposely been omitted, since it will be taken into account later on when the effects of loads which move as well as alternate are being considered. It is sufficient to say at this stage that the general effect of damping is to set a limit to the extent to which synchronous oscillations can mount up and to cause the free oscillations to diminish more or less rapidly, so that, if the applied force persists long enough, only the forced oscillations remain in existence.

CHAPTER IV

OSCILLATIONS PRODUCED BY MOVING LOADS OF CONSTANT MAGNITUDE

Consider the state of oscillation generated in a girder of uniform mass and section freely supported at its ends when subjected to a concentrated force W which moves at a constant speed v in the manner indicated by Fig. 8.

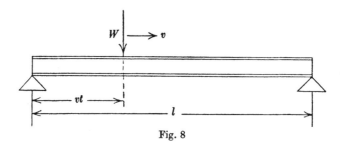

Fig. 8

At time t the force is at a distance vt from the left support and the load distribution at this instant can be expressed by the harmonic series

$$\frac{2W}{l}\left[\sin\frac{\pi vt}{l}\sin\frac{\pi x}{l}+\sin\frac{2\pi vt}{l}\sin\frac{2\pi x}{l}+\sin\frac{3\pi vt}{l}\sin\frac{3\pi x}{l}+\dots\right].$$

Putting $\dfrac{v}{2l}=n$, the series takes the form

$$\frac{2W}{l}\left[\sin\frac{\pi x}{l}\sin 2\pi nt+\sin\frac{2\pi x}{l}\sin 4\pi nt+\sin\frac{3\pi x}{l}\sin 6\pi nt+\dots\right],$$

and harmonic analysis thus reveals the fact that a steady concentrated load is in effect equivalent to a series of stationary but alternating load distributions of a sinusoidal character.

Leaving damping influences out of account, the state of oscillation set up is given by the differential equation

$$EI\frac{d^4y}{dx^4}+m\frac{d^2y}{dt^2}$$
$$=\frac{2W}{l}\left[\sin\frac{\pi x}{l}\sin 2\pi nt+\sin\frac{2\pi x}{l}\sin 4\pi nt+\sin\frac{3\pi x}{l}\sin 6\pi nt+\dots\right].$$

This equation is similar in type to cases dealt with in the previous Chapter, and the forced oscillation or particular integral of this equation takes the form

$$y = \frac{2Wl^3}{\pi^4 EI}\left[\frac{\sin\frac{\pi x}{l}}{1-\left(\frac{n}{n_0}\right)^2}\sin 2\pi nt + \frac{\sin\frac{2\pi x}{l}}{2^4-\left(\frac{2n}{n_0}\right)^2}\sin 4\pi nt + \frac{\sin\frac{3\pi x}{l}}{3^4-\left(\frac{3n}{n_0}\right)^2}\sin 6\pi nt + \ldots\right].$$

In the case of railway bridges, the speed of the moving loads is such that $\frac{n}{n_0}$ is invariably a small fraction.

For example, if $l = 120$ feet and v is taken to have the exceptionally high value 120 feet per sec., $n = \frac{1}{2}$. For such a span n_0 is of the order 6. Consequently, $\frac{n}{n_0} = \frac{1}{12}$, and $\left(\frac{n}{n_0}\right)^2$ is very small in comparison with unity. Accordingly, a very close approximation to the forced oscillation is given by

$$y = \frac{2Wl^3}{\pi^4 EI}\left[\sin\frac{\pi x}{l}\sin 2\pi nt + \frac{1}{2^4}\sin\frac{2\pi x}{l}\sin 4\pi nt + \frac{1}{3^4}\sin\frac{3\pi x}{l}\sin 6\pi nt + \ldots\right],$$

and since $2\pi nt = \frac{\pi vt}{l} = \frac{\pi a}{l}$, where a is the momentary distance which W has moved along the span, the expression above can be written

$$y = \frac{2Wl^3}{\pi^4 EI}\left[\sin\frac{\pi a}{l}\sin\frac{\pi x}{l} + \frac{1}{2^4}\sin\frac{2\pi a}{l}\sin\frac{2\pi x}{l} + \frac{1}{3^4}\sin\frac{3\pi a}{l}\sin\frac{3\pi x}{l} + \ldots\right],$$

and this we have seen in the statical deflection of the girder for this particular position of the load.

Hence to a high degree of approximation (the error hardly amounting to 1 per cent.) the forced oscillation due to the moving load is merely the "crawl-deflection", that is the deflection produced by the load as it moves very slowly along the bridge. Such oscillations as are produced consist of free oscillations which are stimulated into existence at the beginning of the motion to satisfy the condition that the girder shall commence its motion with zero velocity. Introducing the free oscillations necessary to satisfy this condition, the complete solution of the differential equation giving the state of oscillation is

$$y = \frac{2Wl^3}{\pi^4 EI}\left[\frac{\sin\frac{\pi x}{l}}{1-\left(\frac{n}{n_0}\right)^2}\left\{\sin 2\pi nt - \frac{n}{n_0}\sin 2\pi n_0 t\right\}\right.$$

$$\left. + \frac{\sin\frac{2\pi x}{l}}{2^4-\left(\frac{2n}{n_0}\right)^2}\left\{\sin 4\pi nt - \frac{n}{2n_0}\sin 2^2(2\pi n_0 t)\right\} + \ldots\right],$$

and in so far as $\left(\dfrac{n}{n_0}\right)^2$ is negligible in comparison with unity, the motion may be viewed as the "crawl-deflection" upon which is superposed a series of free oscillations given by

$$y = -\frac{2Wl^3}{\pi^4 EI} \cdot \frac{n}{n_0}\left[\sin 2\pi n_0 t \sin\frac{\pi x}{l} + \frac{1}{2^5}\sin 2^2\left(2\pi n_0 t\right)\sin\frac{2\pi x}{l} + \dots\right].$$

The only appreciable dynamic effect is the fundamental free oscillation

$$y = -\frac{2Wl^3}{\pi^4 EI} \cdot \frac{n}{n_0}\sin 2\pi n_0 t \sin\frac{\pi x}{l},$$

and this can be written

$$y = -\frac{n}{n_0} D_W \sin 2\pi n_0 t \sin\frac{\pi x}{l},$$

where D_W is the steady central deflection due to a steady central load W.

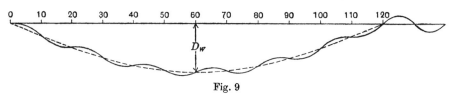

Fig. 9

We have seen that, owing to practical limitations of speed, $\dfrac{n}{n_0}$ is a small fraction which can hardly exceed 1/10. Under these circumstances the dynamic effect of a smooth-running concentrated load will not at its worst add more than 10 per cent. to the maximum "crawl-deflection", and, if railway bridges had only to cater for loads of this character, impact allowance would be a matter of small importance.

The relative insignificance of the dynamical effect of a smooth-running load is illustrated by Fig. 9.

The graph shews the connection between the central deflection of the girder and the position of the load on the bridge. It is drawn for the case where $l = 120$ feet, $v = 120$ f.s. and $n_0 = 6$. It should be observed how the free oscillation eliminates the initial downward slope of the "crawl-deflection", and generates a residual oscillation.

Now consider the case where a uniformly distributed load w per unit length is advancing along a girder at a constant speed v until it covers the whole span. When the head of the load has reached a distance vt as shown in Fig. 10, the load on the span at this instant can be represented by the harmonic series

$$\frac{4w}{\pi}\left[\sin^2\frac{\pi vt}{2l}\sin\frac{\pi x}{l} + \frac{1}{2}\sin^2\frac{2\pi vt}{2l}\sin\frac{2\pi x}{l} + \frac{1}{3}\sin^2\frac{3\pi vt}{2l}\sin\frac{3\pi x}{l} + \dots\right].$$

This series is obtained by putting $b = vt$, and $a = 0$ in a general result established in Chapter I.

The corresponding differential for the equation of motion of the girder is

$$EI\frac{d^4y}{dx^4} + m\frac{d^2y}{dt^2} = \frac{4w}{\pi}\left[\sin^2\pi nt \sin\frac{\pi x}{l} + \frac{1}{2}\sin^2 2\pi nt \sin\frac{2\pi x}{l} + \ldots\right],$$

where $n = \dfrac{v}{2l}$. This can be written

$$EI\frac{d^4y}{dx^4} + m\frac{d^2y}{dt^2} = \frac{2w}{\pi}\left[(1-\cos 2\pi nt)\sin\frac{\pi x}{l} + \frac{1}{2}(1-\cos 4\pi nt)\sin\frac{2\pi x}{l} + \ldots\right],$$

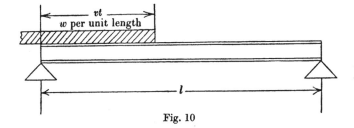

Fig. 10

and the forced oscillation or particular integral is

$$y = \frac{2wl^4}{\pi^5 EI}\left[\sin\frac{\pi x}{l} + \frac{1}{2^5}\sin\frac{2\pi x}{l} + \ldots\right]$$

$$-\frac{2wl^4}{\pi^5 EI}\left[\frac{\sin\dfrac{\pi x}{l}}{1-\left(\dfrac{n}{n_0}\right)^2}\cos 2\pi nt + \frac{1}{2}\frac{\sin\dfrac{2\pi x}{l}}{2^4-\left(\dfrac{2n}{n_0}\right)^2}\cos 4\pi nt + \ldots\right].$$

Neglecting $\left(\dfrac{n}{n_0}\right)^2$ in comparison with unity, the forced oscillation reduces to

$$y = \frac{4wl^4}{\pi^5 EI}\left[\sin^2\pi nt \sin\frac{\pi x}{l} + \frac{1}{2^5}\sin^2 2\pi nt \sin\frac{2\pi x}{l} + \frac{1}{3^5}\sin^2 3\pi nt \sin\frac{3\pi x}{l} + \ldots\right],$$

and, to this degree of approximation, the forced oscillation and the "crawl-deflection" are identical. Moreover, since the approximate expression for y satisfies the condition that $y = 0$ and $\dfrac{dy}{dt} = 0$ when $t = 0$, no free oscillation is generated at the start as was the case when the moving load was concentrated.

For the case of the uniformly distributed advancing load the movement of the girder is almost absolutely free from oscillation. The only departure from accuracy in the foregoing approximation lies in the fact that $\left(\dfrac{n}{n_0}\right)^2$ has been neglected in comparison with unity. The correction necessary to account for this error would add to the forced oscillation an amount given approximately by

$$y = -\frac{2wl^4}{\pi^5 EI}\left(\frac{n}{n_0}\right)^2 \cos 2\pi nt \sin \frac{\pi x}{l},$$

and this in turn necessitates the introduction of a free oscillation having the equation

$$y = \frac{2wl^4}{\pi^5 EI}\left(\frac{n}{n_0}\right)^2 \cos 2\pi n_0 t \sin \frac{\pi x}{l}.$$

The semi-amplitude of this latter can hardly amount to more than half a per cent. of the maximum "crawl-deflection".

CHAPTER V

OSCILLATIONS PRODUCED BY A MOVING ALTERNATING FORCE

Let $P \sin 2\pi Nt$ be this alternating force moving at a constant speed v along the girder so that at a time t it is at a distance vt from the left support, as shewn in Fig. 11.

At this instant the force can be represented by the harmonic load distribution

$$\frac{2P}{l}\left[\sin\frac{\pi x}{l}\sin 2\pi nt + \sin\frac{2\pi x}{l}\sin 4\pi nt + \sin\frac{3\pi x}{l}\sin 6\pi nt + \dots\right]\sin 2\pi Nt,$$

where $n = \dfrac{v}{2l}$.

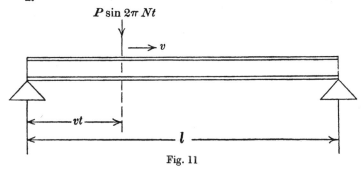

$$P \sin 2\pi Nt$$

Fig. 11

Neglecting damping, the differential equation for the state of oscillation is

$$EI\frac{d^4y}{dx^4} + m\frac{d^2y}{dt^2} = \frac{P}{l}\left[\sin\frac{\pi x}{l}\{\cos 2\pi(N-n)t - \cos 2\pi(N+n)t\}\right.$$
$$\left. + \sin\frac{2\pi x}{l}\{\cos 2\pi(N-2n)t - \cos 2\pi(N+2n)t\} + \dots\right],$$

and the forced oscillation or particular integral is

$$y = \frac{Pl^3}{\pi^4 EI}\left[\sin\frac{\pi x}{l}\left\{\frac{\cos 2\pi(N-n)t}{1-\left(\dfrac{N-n}{n_0}\right)^2} - \frac{\cos 2\pi(N+n)t}{1-\left(\dfrac{N+n}{n_0}\right)^2}\right\}\right.$$
$$\left. + \sin\frac{2\pi x}{l}\left\{\frac{\cos 2\pi(N-2n)t}{2^4-\left(\dfrac{N-2n}{n_0}\right)^2} - \frac{\cos 2\pi(N+2n)t}{2^4-\left(\dfrac{N+2n}{n_0}\right)^2}\right\} + \dots\right].$$

The predominant part of this forced oscillation is usually the primary component, and writing $\dfrac{Pl^3}{\pi^4 EI} = \tfrac{1}{2}D_P$, where D_P is the central deflection due to a steady central load P, this primary component takes the form

$$y = \frac{D_P}{2}\left[\frac{\cos 2\pi\,(N-n)\,t}{1-\left(\dfrac{N-n}{n_0}\right)^2} - \frac{\cos 2\pi\,(N+n)\,t}{1-\left(\dfrac{N+n}{n_0}\right)^2}\right]\sin\frac{\pi x}{l},$$

and, to this degree of approximation, the effect of the moving alternating force is identical with that of two stationary central alternating forces of magnitude $\dfrac{P}{2}\cos 2\pi\,(N-n)\,t$ and $-\dfrac{P}{2}\cos 2\pi\,(N+n)\,t$.

To adjust the condition that $y=0$ when $t=0$, a free oscillation of the type

$$y = A\cos 2\pi n_0 t\sin\frac{\pi x}{l}$$

must be superposed, and as a result the first and usually predominant component of the complete oscillation is

$$y = \frac{D_P}{2}\left[\frac{\cos 2\pi\,(N-n)\,t-\cos 2\pi n_0 t}{1-\left(\dfrac{N-n}{n_0}\right)^2} - \frac{\cos 2\pi\,(N+n)\,t-\cos 2\pi n_0 t}{1-\left(\dfrac{N+n}{n_0}\right)^2}\right]\sin\frac{\pi x}{l}.$$

The second harmonic component is

$$y = \frac{D_P}{2}\left[\frac{\cos 2\pi\,(N-2n)\,t-\cos 2^2\,(2\pi n_0 t)}{2^4-\left(\dfrac{N-2n}{n_0}\right)^2}\right.$$
$$\left. - \frac{\cos 2\pi\,(N+2n)\,t-\cos 2^2\,(2\pi n_0 t)}{2^4-\left(\dfrac{N+2n}{n_0}\right)^2}\right]\sin\frac{2\pi x}{l},$$

and by inspection the higher harmonic components can be written down in a similar form.

The case when $N=n_0$ demands special attention, since in this case the most violent state of oscillation is developed. The frequency of the alternating force then agrees with the fundamental frequency of the girder; the oscillations become resonant in character and continue to increase up to the instant when the alternating force passes off the girder. Under these circumstances the higher harmonic components shrink into relative insignificance and the state of oscillation when $N=n_0$ is very accurately given by

$$y = \frac{D_P}{2}\left[\frac{\cos 2\pi\,(N-n)\,t-\cos 2\pi N t}{1-\left(1-\dfrac{n}{N}\right)^2} - \frac{\cos 2\pi\,(N+n)\,t-\cos 2\pi N t}{1-\left(1+\dfrac{n}{N}\right)^2}\right]\sin\frac{\pi x}{l},$$

i.e. $y = \dfrac{D_P N^2}{2} \left[\dfrac{\cos 2\pi (N-n)t - \cos 2\pi Nt}{n(2N-n)} \right.$

$\left. + \dfrac{\cos 2\pi (N+n)t - \cos 2\pi Nt}{n(2N+n)} \right] \sin \dfrac{\pi x}{l},$

i.e. $y = \dfrac{D_P N^2}{2n} \left[\cos 2\pi Nt (\cos 2\pi nt - 1) \left(\dfrac{1}{2N-n} + \dfrac{1}{2N+n} \right) \right.$

$\left. + \sin 2\pi Nt \sin 2\pi nt \left(\dfrac{1}{2N-n} - \dfrac{1}{2N+n} \right) \right] \sin \dfrac{\pi x}{l},$

i.e. $y = \dfrac{D_P N^2}{2n(4N^2 - n^2)} [4N \cos 2\pi Nt (\cos 2\pi nt - 1) + 2n \sin 2\pi Nt \sin 2\pi nt] \sin \dfrac{\pi x}{l},$

i.e. $y = \dfrac{D_P N^2}{4N^2 - n^2} \left[\sin 2\pi nt \sin 2\pi Nt - \dfrac{2N}{n}(1 - \cos 2\pi nt) \cos 2\pi Nt \right] \sin \dfrac{\pi x}{l}.$

In the application of this theory to railway bridges, the alternating force is due to the balance-weights attached to the driving-wheels and N, accordingly, is the number of revolutions made by the wheels per sec. In such cases

$$\frac{n}{N} = \frac{v}{2l} \times \frac{\text{circumference of wheel}}{v} = \frac{\text{circumference of wheel}}{\text{twice span of bridge}},$$

and, except in the case of very short bridges, $\dfrac{n}{2N}$ is a small fraction. Assuming this to be the case, an excellent approximation to the state of oscillation set up by the moving synchronous alternating force is given by

$$y = -\frac{D_P}{2} \cdot \frac{N}{n} (1 - \cos 2\pi nt) \cos 2\pi Nt \sin \frac{\pi x}{l}.$$

The oscillations starting from rest mount up progressively as the alternating force passes along the bridge. At the instant the force is leaving the bridge $2\pi nt = \pi$ and the dynamic magnification is $\dfrac{N}{n}$. At the instant the force has reached the centre of the bridge $2\pi nt = \dfrac{\pi}{2}$ and the dynamic magnification is $\dfrac{N}{2n}$, that is to say, it is given by the number of alternations made by the force as it traverses the span.

To illustrate the foregoing theory, consider the following example:

Suppose a bridge of 270 feet span, idealized to be of uniform mass and section, has a fundamental natural frequency of 3 and a total mass of 450 tons. Suppose that it is traversed by a locomotive which weighs 100 tons and that the balance-weights on its driving-wheels set up an alternating rail pressure of magnitude $0 \cdot 6 N^2 \sin 2\pi Nt$ tons, where N denotes the number of wheel revolutions per sec.

In order to fit in with the foregoing theory, the action of the locomotive on the bridge will be idealized into a concentrated force of 100 tons combined with an alternating force of $0\cdot6N^2\sin 2\pi Nt$ tons, moving together at a constant speed along the bridge. The inertia effect of the locomotive and damping influences will be ignored.

The most violent state of oscillation will be considered, namely that which is set up when the speed of the locomotive is such that $N=3$ and consequently a state of synchronism exists between the frequency of the alternating force and the fundamental frequency of the bridge.

At this speed the alternating force is $5\cdot4\sin 6\pi t$ tons and, if the circumference of the driving-wheels is 15 feet, $v=45$ f.s., $n=\dfrac{v}{2l}=\dfrac{1}{12}$ and $\dfrac{N}{n}=36$. Inserting these values in the general expression for the oscillation, viz.

$$y=-\frac{D_P}{2}\cdot\frac{N}{n}(1-\cos 2\pi nt)\cos 2\pi Nt\sin\frac{\pi x}{l},$$

the oscillation for this particular case takes the form

$$y=-18D_P\left(1-\cos\frac{\pi t}{6}\right)\cos 6\pi t\sin\frac{\pi x}{l}.$$

Let D_W be the central deflection due to the load of 100 tons standing at the centre of the bridge. Then $D_P=\dfrac{5\cdot4}{100}D_W$ and the equation for the oscillation produced by the moving alternating force takes the form

$$y=-0\cdot972D_W\left(1-\cos\frac{\pi t}{6}\right)\cos 6\pi t\sin\frac{\pi x}{l}.$$

Upon this, the state of oscillation set up by the moving force of 100 tons must be superposed. It was shewn in Chapter IV that this can be viewed as the "crawl-deflection" combined with a fundamental free oscillation given by

$$y=-D_W\frac{n}{n_0}\sin 2\pi n_0 t\sin\frac{\pi x}{l},$$

which, for this particular case, takes the form

$$y=-0\cdot028D_W\sin 6\pi t\sin\frac{\pi x}{l}.$$

Hence the complete oscillation produced by the joint action of the constant force of 100 tons, and the alternating force of $5\cdot4\sin 6\pi t$ tons, can be viewed as the "crawl-deflection" for the constant force combined with an oscillation having the equation

$$y=-D_W\left[0\cdot972\left(1-\cos\frac{\pi t}{6}\right)\cos 6\pi t+0\cdot028\sin 6\pi t\right]\sin\frac{\pi x}{l}.$$

If a is the distance the combined forces have moved along the bridge in time t, then $a = vt$ and $\dfrac{\pi t}{6} = \dfrac{\pi a}{l}$.

Hence for the oscillation which has to be added to the "crawl-deflection", the connection between δ, the central deflection, and a, the position of the load, is given by

$$\delta = - D_W \left[0 \cdot 972 \left(1 - \cos \frac{\pi a}{l} \right) \cos 6\pi t + 0 \cdot 028 \sin 6\pi t \right].$$

The central deflection for the "crawl-deflection" is given by

$$\delta = D_W \sin \frac{\pi a}{l}.$$

The result of superposing these two values of δ is illustrated by Fig. 12,

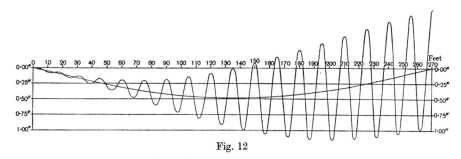

Fig. 12

which shews the value of the central deflection of the bridge in terms of the position of the moving load.

The value of D_W in the foregoing case, as determined by the formula $D_W = \dfrac{W}{2\pi^2 n_0{}^2 M_G}$, is $0 \cdot 483$ in.

Due to the oscillation the maximum deflection is approximately $2 \cdot 4 D_W = 1 \cdot 16$ ins., and this dynamic increment of 140 per cent. is considerably in excess of that found in practice with railway bridges of a similar type. The exaggeration is due to the over-idealization which has been employed to bring the problem within the scope of the limited analysis so far developed. This idealization excludes damping; it also leaves out of account the inertia effects of the locomotive and the consequent change in the natural frequency of the bridge as the locomotive moves along the span. To bring theory and experiment into agreement these omissions must be rectified, and this is done in the more comprehensive method of analysis which is developed in the next three Chapters.

CHAPTER VI

OSCILLATIONS PRODUCED BY A MOVING ALTERNATING FORCE, TAKING INTO ACCOUNT DAMPING

As in the previous Chapter, let $P \sin 2\pi N t$ be an alternating force moving at a constant speed v along a girder of uniform mass and section freely supported at its ends.

Following the notation adopted in Chapter II, the damping influence will be taken to consist of a resistance to vertical motion amounting to $4\pi n_b m \dfrac{dy}{dt}$ per unit length distributed along the girder.

Expressing the alternating force by the now familiar harmonic series, the equation of motion for the state of oscillation is

$$EI\frac{d^4y}{dx^4} + 4\pi n_b m \frac{dy}{dt} + m \frac{d^2y}{dt^2}$$
$$= \frac{2P}{l}\left[\sin 2\pi nt \sin \frac{\pi x}{l} + \sin 4\pi nt \sin \frac{2\pi x}{l} + \dots\right]\sin 2\pi Nt,$$

where $n = \dfrac{v}{2l}$, i.e.

$$EI\frac{d^4y}{dx^4} + 4\pi n_b m \frac{dy}{dt} + m \frac{d^2y}{dt^2} = \frac{P}{l}\left[\{\cos 2\pi(N-n)t - \cos 2\pi(N+n)t\}\sin \frac{\pi x}{l}\right.$$
$$\left. + \{\cos 2\pi(N-2n)t - \cos 2\pi(N+2n)t\}\sin \frac{2\pi x}{l} + \dots\right].$$

For the primary component of the forced oscillation assume

$$y = f(t)\sin\frac{\pi x}{l}.$$

The equation for $f(t)$ is accordingly

$$\frac{d^2f}{dt^2} + 4\pi n_b \frac{df}{dt} + 4\pi^2 n_0^2 f(t) = \frac{P}{M_G}[\cos 2\pi(N-n)t - \cos 2\pi(N+n)t],$$

where M_G is the total mass of the girder. Hence

$$f(t) = \frac{P}{4\pi^2 n_0^2 M_G}\left[\frac{\cos\{2\pi(N-n)t - \Phi\}}{\left\{\left[1 - \left(\frac{N-n}{n_0}\right)^2\right]^2 + \left[\frac{2n_b(N-n)}{n_0^2}\right]^2\right\}^{\frac{1}{2}}}\right.$$
$$\left. - \frac{\cos\{2\pi(N+n)t - \Psi\}}{\left\{\left[1 - \left(\frac{N+n}{n_0}\right)^2\right]^2 + \left[\frac{2n_b(N+n)}{n_0^2}\right]^2\right\}^{\frac{1}{2}}}\right],$$

where $\qquad \tan \Phi = \dfrac{2n_b\,(N-n)}{n_0{}^2 - (N-n)^2} \quad$ and $\quad \tan \Psi = \dfrac{2n_b\,(N+n)}{n_0{}^2 - (N+n)^2}.$

It was shewn in Chapter II that D_P, the central deflection due to a steady central load P, is given by

$$D_P = \frac{P}{2\pi^2 n_0{}^2 M_G}.$$

Accordingly, the primary component of the forced oscillation can be written in the form

$$y = \frac{D_P}{2}\left[\frac{\cos\left\{2\pi\,(N-n)\,t - \Phi\right\}}{\left\{\left[1 - \left(\dfrac{N-n}{n_0}\right)^2\right]^2 + \left[\dfrac{2n_b\,(N-n)}{n_0{}^2}\right]^2\right\}^{\frac{1}{2}}} \right.$$

$$\left. - \frac{\cos\left\{2\pi\,(N+n)\,t - \Psi\right\}}{\left\{\left[1 - \left(\dfrac{N+n}{n_0}\right)^2\right]^2 + \left[\dfrac{2n_b\,(N+n)}{n_0{}^2}\right]^2\right\}^{\frac{1}{2}}} \right] \sin\frac{\pi x}{l}.$$

In a similar manner the pth harmonic component of the forced oscillation can be deduced in the form

$$y = \frac{D_P}{2}\left[\frac{\cos\left\{2\pi\,(N-pn)\,t - \Phi_p\right\}}{\left\{\left[p^4 - \left(\dfrac{N-pn}{n_0}\right)^2\right]^2 + \left[\dfrac{2n_b\,(N-pn)}{n_0{}^2}\right]^2\right\}^{\frac{1}{2}}} \right.$$

$$\left. - \frac{\cos\left\{2\pi\,(N+pn)\,t - \Psi_p\right\}}{\left\{\left[p^4 - \left(\dfrac{N+pn}{n_0}\right)^2\right]^2 + \left[\dfrac{2n_b\,(N+pn)}{n_0{}^2}\right]^2\right\}^{\frac{1}{2}}} \right] \sin\frac{p\pi x}{l},$$

where $\quad \tan \Phi_p = \dfrac{2n_b\,(N-pn)}{p^4 n_0{}^2 - (N-pn)^2} \quad$ and $\quad \tan \Psi_p = \dfrac{2n_b\,(N+pn)}{p^4 n_0{}^2 - (N+pn)^2}.$

In general, and particularly for the synchronous case when $N = n_0$, these higher harmonic components of the forced oscillation are relatively insignificant in comparison with the first.

Since the first harmonic component does not satisfy either of the starting conditions which require that $y = 0$ and also $\dfrac{dy}{dt} = 0$ when $t = 0$, a free oscillation of the type

$$y = e^{-2\pi n_b t}\left[A \sin 2\pi n_0 t + B \cos 2\pi n_0 t\right] \sin\frac{\pi x}{l}$$

is called into existence.

Adjusting the constants A and B to satisfy these starting conditions, the primary component of the oscillation takes the form

$$
\begin{aligned}
y = \frac{D_P}{2} & \left[\frac{\cos\{2\pi(N-n)t-\Phi\} - e^{-2\pi n_b t}\left\{\left(\dfrac{N-n}{n_0}\sin\Phi + \dfrac{n_b}{n_0}\cos\Phi\right) \times \sin 2\pi n_0 t + \cos\Phi\cos 2\pi n_0 t\right\}}{\left\{\left[1-\left(\dfrac{N-n}{n_0}\right)^2\right]^2 + \left[\dfrac{2n_b(N-n)}{n_0{}^2}\right]^2\right\}^{\frac{1}{2}}} \right] \sin\frac{\pi x}{l} \\[2em]
- \frac{D_P}{2} & \left[\frac{\cos\{2\pi(N+n)t-\Psi\} - e^{-2\pi n_b t}\left\{\left(\dfrac{N+n}{n_0}\sin\Psi + \dfrac{n_b}{n_0}\cos\Psi\right) \times \sin 2\pi n_0 t + \cos\Psi\cos 2\pi n_0 t\right\}}{\left\{\left[1-\left(\dfrac{N+n}{n_0}\right)^2\right]^2 + \left[\dfrac{2n_b(N+n)}{n_0{}^2}\right]^2\right\}^{\frac{1}{2}}} \right] \sin\frac{\pi x}{l}.
\end{aligned}
$$

Consider next how the state of oscillation set up by a constant force W moving along a girder at a uniform speed, which was dealt with in Chapter IV, is modified by the influence of damping.

The differential equation for the motion there given is only changed by the introduction of the damping term $4\pi n_b m\dfrac{dy}{dt}$ on the left side of the equation, which accordingly reads

$$
EI\frac{d^4y}{dx^4} + 4\pi n_b m\frac{dy}{dt} + m\frac{d^2y}{d^2t} = \frac{2W}{l}\left[\sin\frac{\pi x}{l}\sin 2\pi nt + \sin\frac{2\pi x}{l}\sin 4\pi nt + \dots\right],
$$

and the corresponding forced oscillation is

$$
y = D_W\left[\frac{\sin(2\pi nt-\phi_1)\sin\dfrac{\pi x}{l}}{\left\{\left[1-\left(\dfrac{n}{n_0}\right)^2\right]^2 + \left[\dfrac{2n_b n}{n_0{}^2}\right]^2\right\}^{\frac{1}{2}}} + \frac{\sin(4\pi nt-\phi_2)\sin\dfrac{2\pi x}{l}}{\left\{\left[2^4-\left(\dfrac{2n}{n_0}\right)^2\right]^2 + \left[\dfrac{4n_b n}{n_0{}^2}\right]^2\right\}^{\frac{1}{2}}} + \dots \right],
$$

where
$$
\tan\phi_1 = \frac{2n_b n}{n_0{}^2-n^2}, \quad \tan\phi_2 = \frac{4n_b n}{2^4 n_0{}^2-(2n)^2}
$$

and D_W is the central deflection due to a steady central load W.

In the application of this theory to railway bridges n and n_b are so small in comparison with n_0 that ϕ_1 and ϕ_2 are hardly perceptible, and the forced oscillation is given to a high degree of accuracy by the equation

$$
y = D_W\left[\sin\frac{\pi x}{l}\sin 2\pi nt + \frac{1}{2^4}\sin\frac{2\pi x}{l}\sin 4\pi nt + \dots\right].
$$

This is equivalent to saying that the forced oscillation is indistinguishable from the "crawl-deflection".

To satisfy the starting conditions free oscillations of the type

$$y = e^{-2\pi n_b t} A_1 \sin 2\pi n_0 t \sin \frac{\pi x}{l},$$

$$y = e^{-2\pi n_b t} A_2 \sin 2^2 (2\pi n_0 t) \sin \frac{\pi x}{l}, \text{ etc.}$$

must be introduced.

Adjusting the constants of integration to satisfy the starting conditions, the free oscillation which has to be superposed on the "crawl-deflection" is given by

$$y = -\frac{n}{n_0} D_W e^{-2\pi n_b t} \left[\sin \frac{\pi x}{l} \sin 2\pi n_0 t + \frac{1}{2^4} \sin \frac{2\pi x}{l} \sin 2^2 (2\pi n_0 t) + \dots \right].$$

The only part of this which is of any practical importance is the primary component

$$y = -\frac{n}{n_0} D_W e^{-2\pi n_b t} \sin \frac{\pi x}{l} \sin 2\pi n_0 t.$$

This, even at the start, is quite a small oscillation, and it dies out as the load passes along the bridge, the common ratio of successive oscillations being $e^{-2\pi \frac{n_b}{n_0}}$.

FORCED OSCILLATION FOR THE SYNCHRONOUS CASE
WHEN $N = n_0$

Putting $N = n_0$ in the expression for the forced oscillation previously established, the oscillation for this particular case takes the form

$$y = \frac{P}{4\pi^2 M_G} \left[\frac{\cos\{2\pi(N-n)t - \Phi\}}{\{n^2(2N-n)^2 + 4n_b^2(N-n)^2\}^{\frac{1}{2}}} - \frac{\cos\{2\pi(N+n)t - \Psi'\}}{\{n^2(2N+n)^2 + 4n_b^2(N+n)^2\}^{\frac{1}{2}}} \right] \sin \frac{\pi x}{l}.$$

Write this

$$y = \frac{P}{4\pi^2 M_G} \left[\frac{\cos\{2\pi(N-n)t - \Phi\}}{D_{-1}} - \frac{\cos\{2\pi(N+n)t - \Psi'\}}{D_1} \right] \sin \frac{\pi x}{l},$$

i.e.

$$y = \frac{P}{4\pi^2 M_G} \left[\left\{ \cos 2\pi n t \left(\frac{\cos \Phi}{D_{-1}} - \frac{\cos \Psi'}{D_1} \right) - \sin 2\pi n t \left(\frac{\sin \Phi}{D_{-1}} + \frac{\sin \Psi'}{D_1} \right) \right\} \cos 2\pi N t \right.$$

$$\left. + \left\{ \cos 2\pi n t \left(\frac{\sin \Phi}{D_{-1}} - \frac{\sin \Psi'}{D_1} \right) + \sin 2\pi n t \left(\frac{\cos \Phi}{D_{-1}} + \frac{\cos \Psi'}{D_1} \right) \right\} \sin 2\pi N t \right] \sin \frac{\pi x}{l}.$$

In this expression

$$\sin\Phi = \frac{2n_b(N-n)}{D_{-1}}, \quad \cos\Phi = \frac{n(2N-n)}{D_{-1}},$$

$$\sin\Psi = \frac{2n_b(N+n)}{D_1}, \quad \cos\Psi = -\frac{n(2N+n)}{D_1}.$$

On the assumption, which is justifiable in all practical cases, that n^2 and n_b^2 are negligible in comparison with unity,

$$\frac{\cos\Phi}{D_{-1}} - \frac{\cos\Psi}{D_1} = \frac{n}{N}\cdot\frac{1}{n^2+n_b^2}, \quad \frac{\sin\Phi}{D_{-1}} + \frac{\sin\Psi}{D_1} = \frac{n_b}{N}\cdot\frac{1}{n^2+n_b^2},$$

$$\frac{\cos\Phi}{D_{-1}} + \frac{\cos\Psi}{D_1} = \frac{n^2}{2N^2}\cdot\frac{n^2+3n_b^2}{(n^2+n_b^2)^2}, \quad \frac{\sin\Phi}{D_{-1}} - \frac{\sin\Psi}{D_1} = \frac{nn_b^3}{N^2(n^2+n_b^2)^2},$$

and, to this very close degree of approximation, the equation for the forced oscillation takes the form

$$y = \frac{P}{4\pi^2 M_G}\left[\frac{1}{N(n^2+n_b^2)}\{n\cos 2\pi nt - n_b\sin 2\pi nt\}\cos 2\pi Nt \right.$$

$$\left. + \frac{1}{N^2(n^2+n_b^2)^2}\left\{nn_b^3\cos 2\pi nt + \frac{n^2}{2}(n^2+3n_b^2)\sin 2\pi nt\right\}\sin 2\pi Nt\right]\sin\frac{\pi x}{l}.$$

In this expression, the component associated with $\cos 2\pi Nt$ predominates to such an extent, that the forced oscillation is almost exactly in quadrature with the applied alternating force and is very nearly represented by

$$y = \frac{P}{4\pi^2 N^2 M_G}\cdot\frac{N}{n^2+n_b^2}\{n\cos 2\pi nt - n_b\sin 2\pi nt\}\cos 2\pi Nt\sin\frac{\pi x}{l}.$$

The justification of this statement will be apparent in the numerical example by which this theory is illustrated later on in this Chapter.

ADDED FREE OSCILLATION TO SATISFY
THE STARTING CONDITIONS

The approximate expression for the forced oscillation for the synchronous case $N = n_0$ is

$$y = \frac{P}{4\pi^2 N^2 M_G}\cdot\frac{N}{n^2+n_b^2}\{n\cos 2\pi nt - n_b\sin 2\pi nt\}\cos 2\pi Nt\sin\frac{\pi x}{l}.$$

Initially the central deflection δ is given by

$$\delta = \frac{P}{4\pi^2 N^2 M_G}\cdot\frac{Nn}{n^2+n_b^2},$$

and initially

$$\frac{d(\delta)}{dt} = -\frac{P}{4\pi^2 N^2 M_G}\cdot\frac{Nn_b}{n^2+n_b^2}\cdot 2\pi n.$$

Since $N = n_0$, the free oscillation to be superposed has the general form

$$y = e^{-2\pi n_b t} [A \sin 2\pi N t + B \cos 2\pi N t] \sin \frac{\pi x}{l},$$

and the central deflection is accordingly given by

$$\delta = e^{-2\pi n_b t} [A \sin 2\pi N t + B \cos 2\pi N t].$$

Initially $\delta = B$ and $\dfrac{d(\delta)}{dt} = 2\pi N A - 2\pi n_b B$. Hence to satisfy the starting conditions

$$B = -\frac{P}{4\pi^2 N^2 M_G} \cdot \frac{Nn}{n^2 + n_b{}^2},$$

and

$$NA - n_b B = +\frac{P}{4\pi^2 N^2 M_G} \cdot \frac{Nnn_b}{n^2 + n_b{}^2}.$$

Thus $A = 0$, and the free oscillation which has to be superposed on the forced oscillation to satisfy the starting conditions is

$$y = -\frac{P}{4\pi^2 N^2 M_G} \cdot \frac{Nn}{n^2 + n_b{}^2} e^{-2\pi n_b t} \cos 2\pi N t \sin \frac{\pi x}{l}.$$

Hence for the synchronous case ($N = n_0$), the complete state of oscillation built up by an alternating force $P \sin 2\pi N t$ as it moves with velocity v along the bridge is approximately given by

$$y = \frac{P}{4\pi^2 N^2 M_G} \cdot \frac{N}{n^2 + n_b{}^2} [n (\cos 2\pi n t - e^{-2\pi n_b t}) - n_b \sin 2\pi n t] \cos 2\pi N t \sin \frac{\pi x}{l}.$$

Suppose that the alternating force is associated with a steady force W moving with it along the bridge, and suppose that the central deflection produced by W when steadily applied at the centre of the bridge is D_W; then, since $D_W = \dfrac{W}{2\pi^2 N^2 M_G}$, the oscillation produced by the alternating force $P \sin 2\pi N t$ can be written in the form

$$y = \frac{P}{2W} \cdot \frac{N}{n^2 + n_b{}^2} D_W [n (\cos 2\pi n t - e^{-2\pi n_b t}) - n_b \sin 2\pi n t] \cos 2\pi N t \sin \frac{\pi x}{l}.$$

For the particular case $n_b = 0$, this reduces to the approximation deduced in the previous Chapter.

NUMERICAL EXAMPLE

Consider the case dealt with in the previous Chapter. The bridge, idealized to be of uniform mass and section, has a span of 270 feet and a total mass of 450 tons. The effect of a locomotive crossing the bridge is idealized to consist of a gravity force of 100 tons combined with an alternating force due to the balance-weights on the driving-wheels. The natural frequency of the bridge is 3, and when the driving-wheels are making 3 revs. per sec. resonance is established and the magnitude of the alternating force is $5\cdot4\sin6\pi t$ tons. The circumference of the driving-wheels being 15 feet, the speed of the loco-motive is 45 f.s. and $n=\dfrac{v}{2l}=\dfrac{1}{12}$.

For the damping, the value of n_b is taken to be $0\cdot12$. The common ratio for successive free oscillation is $e^{-2\pi\frac{n_b}{n_0}}$, and with the above value of n_b, this ratio has the value $0\cdot78$, which is a suitable value for a bridge of 270 feet span.

With these values of N, n, n_b, M_G and P,

$$\frac{nN}{n^2+n_b{}^2}=11\cdot71,\quad \frac{n_bN}{n^2+n_b{}^2}=16\cdot87,$$

$$\frac{nn_b{}^3}{(n^2+n_b{}^2)^2}=0\cdot32,\quad \frac{n^2(n^2+3n_b{}^2)}{2N^2(n^2+n_b{}^2)^2}=0\cdot38,$$

$$\frac{P}{4\pi^2N^2M_G}=\frac{5\cdot4\times12g}{16200}\ \text{in.}=0\cdot013\,\text{in.},$$

the forced oscillation in inches is given by the expression

$$y=0\cdot013\,[11\cdot17\cos2\pi nt-16\cdot87\sin2\pi nt]\cos2\pi Nt\sin\frac{\pi x}{l}$$

$$+0\cdot013\,[0\cdot32\cos2\pi nt+0\cdot38\sin2\pi nt]\sin2\pi Nt\sin\frac{\pi x}{l},$$

and, as mentioned previously, the component associated with $\sin2\pi Nt$ can be omitted without any serious loss of accuracy.

The steady central deflection due to a steadily applied central load of 100 tons is $0\cdot482$ in. Calling this D_W, the equation of the forced oscillation due to the alternating force is very nearly represented by

$$y=\frac{D_W}{100}\,[31\cdot6\cos2\pi nt-45\cdot5\sin2\pi nt]\cos2\pi Nt\sin\frac{\pi x}{l}.$$

Since $2\pi nt=\dfrac{\pi vt}{l}=\dfrac{\pi a}{l}$, where a is the distance the load has moved along the bridge, the value of the central deflection δ in terms of a for this forced oscillation is given by

$$\delta=\frac{D_W}{100}\left[31\cdot6\cos\frac{\pi a}{l}-45\cdot5\sin\frac{\pi a}{l}\right]\cos6\pi t.$$

The complete expression for the oscillation due to the alternating force, obtained by adding to the forced oscillation a free oscillation to adjust the starting conditions, is obtained by evaluating the general expression

$$y = \frac{P}{2W} \cdot \frac{N}{n^2 + n_b{}^2} D_W \left[n \left(\cos 2\pi n t - e^{-2\pi n_b t} \right) - n_b \sin 2\pi n t \right] \cos 2\pi N t \sin \frac{\pi x}{l}.$$

This yields

$$y = \frac{D_W}{100} \left[31 \cdot 6 \left(\cos \frac{\pi t}{6} - e^{-0 \cdot 754 t} \right) - 45 \cdot 5 \sin \frac{\pi t}{6} \right] \cos 6\pi t \sin \frac{\pi x}{l},$$

and the central deflection δ in terms of a, the distance the load has moved along the bridge, is given by

$$\delta = \frac{D_W}{100} \left[31 \cdot 6 \left(\cos \frac{\pi a}{l} - e^{-0 \cdot 754 \frac{a}{v}} \right) - 45 \cdot 5 \sin \frac{\pi a}{l} \right] \cos 6\pi t.$$

OSCILLATION DUE TO THE STEADY MOVING LOAD

The forced oscillation is very accurately represented by the primary component of the "crawl-deflection", that is to say, by

$$y = D_W \sin 2\pi n t \sin \frac{\pi x}{l}.$$

The free oscillation which has to be superposed to adjust the starting conditions is given by

$$y = -\frac{n}{N} e^{-2\pi n_b t} \sin 2\pi N t \sin \frac{\pi x}{l}.$$

Putting in the values of n, N and n_b, the state of oscillation due to the moving steady load is

$$y = \frac{D_W}{100} \left[100 \sin \frac{\pi t}{6} - 2 \cdot 78 e^{-0 \cdot 754 t} \sin 6\pi t \right] \sin \frac{\pi x}{l},$$

and the central deflection δ in terms of a, the distance the load has moved along the bridge, is given by

$$\delta = \frac{D_W}{100} \left[100 \sin \frac{\pi a}{l} - 2 \cdot 78 e^{-0 \cdot 754 \frac{a}{v}} \sin 6\pi t \right].$$

Fig. 13, which is the result of plotting the connection between δ and a given by this last expression, illustrates the insignificant amount of oscillation developed if the locomotive is free from hammer-blow, and can be viewed as a constant force moving at a constant speed. It confirms what has been stated previously, that for smooth-running loads, such as electric or four-cylinder locomotives, impact allowances are hardly required except so far as they may be necessary to cover the effects of rail joints or lurching due to track imperfections.

Fig. 14 shews the result of superposing the hammer-blow amounting to $5 \cdot 4 \sin 6\pi t$. It should be compared with the similar case illustrated by Fig. 12 in Chapter V, in which the effect of damping was excluded.

It will be seen that the maximum deflection is reduced to about $1 \cdot 5 D_W$, whereas, in the absence of damping, it was approximately $2 \cdot 4 D_W$. A theory which fails to take damping into account is useless for predicting bridge oscillations. A knowledge of the damping characteristics of bridges as revealed by their residual oscillations is most essential and it would be advantageous if more information on this point was available.

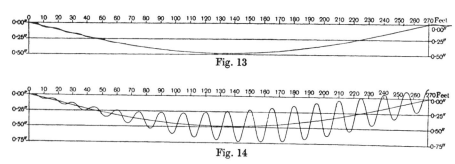

Fig. 13

Fig. 14

In its application to practical cases the foregoing theory is still defective in that it does not take into account the inertia effect of the moving load. This omission is rectified in the more comprehensive analysis developed in the next two Chapters, but, for the benefit of the reader, it should be pointed out that the main object of the somewhat formidable and even repulsive mathematics set forth in Chapters VII and VIII is to set up standards for checking the validity and trustworthiness of comparatively simple formulae, for predicting the most violent state of oscillation to which a given long-span bridge will be subjected, due to the passage of any given locomotive. The establishment of these approximate formulae is based upon the analysis developed in this Chapter, and the reader who is prepared to take on trust the validity of the mathematics and arithmetical computations contained in the next two Chapters can proceed at once to study the comparatively simple approximate process for taking into account the inertia of the moving load, which is explained and justified at the end of Chapter VIII.

CHAPTER VII

OSCILLATIONS PRODUCED BY A MOVING ALTERNATING FORCE ASSOCIATED WITH A CONCENTRATED MOVING MASS

To simplify the mathematical analysis as much as possible, the effect of damping will be omitted for the time being, and the inclusion of this effect will be deferred until Chapter VIII.

The problem to be investigated is illustrated by Fig. 15.

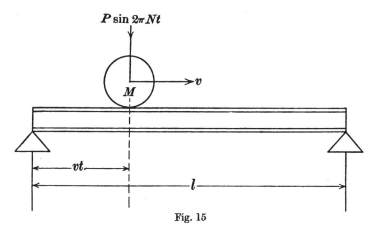

Fig. 15

If at any instant t, α is the downward acceleration of M, the girder will have to support a concentrated downward force $-M\alpha$, owing to the inertia of the load. Leaving out of account gravity forces, the total load concentrated at the section $x = vt$ is $P \sin 2\pi Nt - M\alpha$, and the primary harmonic component of this force is the load distribution

$$\frac{2}{l}(P \sin 2\pi Nt - M\alpha) \sin \frac{\pi vt}{l} \sin \frac{\pi x}{l}.$$

In the absence of damping the primary component of the state of oscillation, which is the only component of practical importance, is given by the equation

$$EI \frac{d^4y}{dx^4} + m \frac{d^2y}{dt^2} = \frac{2}{l}(P \sin 2\pi Nt - M\alpha) \sin \frac{\pi vt}{l} \sin \frac{\pi x}{l}.$$

Let $y = f(t) \sin \frac{\pi x}{l}$ be the solution.

The vertical displacement of M at time t is

$$f(t)\sin\frac{\pi vt}{l}.$$

The downward vertical velocity of M is

$$\frac{\pi v}{l}\cos\frac{\pi vt}{l}f(t)+\sin\frac{\pi vt}{l}\frac{df}{dt}.$$

The downward vertical acceleration of M is

$$-\frac{\pi^2 v^2}{l^2}\sin\frac{\pi vt}{l}f(t)+\frac{2\pi v}{l}\cos\frac{\pi vt}{l}\frac{df}{dt}+\sin\frac{\pi vt}{l}\frac{d^2f}{dt^2}.$$

Putting $n=\dfrac{v}{2l}$, the vertical downward acceleration of M is given by

$$\alpha=-4\pi^2 n^2\sin 2\pi nt\,f(t)+4\pi n\cos 2\pi nt\,\frac{df}{dt}+\sin 2\pi nt\,\frac{d^2f}{dt^2}.$$

Inserting this value of α, the equation for the oscillation given above takes the form

$$EI\frac{d^4y}{dx^4}+m\frac{d^2y}{dt^2}=\frac{2}{l}\left[P\sin 2\pi Nt-M\left\{-4\pi^2 n^2\sin 2\pi nt\,f(t)\right.\right.$$
$$\left.\left.+4\pi n\cos 2\pi nt\,\frac{df}{dt}+\sin 2\pi nt\,\frac{d^2f}{dt^2}\right\}\right]\sin 2\pi nt\sin\frac{\pi x}{l}$$

and $y=f(t)\sin\dfrac{\pi x}{t}$ fits this equation if

$$EI\frac{\pi^4}{l^4}f(t)+m\frac{d^2f}{dt^2}=\frac{2}{l}\left[P\sin 2\pi Nt-M\left\{-4\pi^2 n^2\sin 2\pi nt\,f(t)\right.\right.$$
$$\left.\left.+4\pi n\cos 2\pi nt\,\frac{df}{dt}+\sin 2\pi nt\,\frac{d^2f}{dt^2}\right\}\right]\sin 2\pi nt.$$

Putting $EI\dfrac{\pi^4}{l^4}=4\pi^2 n_0^2 m$, where n_0 is the unloaded fundamental frequency of the girder, and $ml=M_G$, where M_G is the total mass of the girder, the equation for $f(t)$ takes the form

$$4\pi^2\left(n_0^2-\frac{2M}{M_G}n^2\sin 2\pi nt\right)f(t)+4\pi n\frac{M}{M_G}\sin 4\pi nt\,\frac{df}{dt}+\left(1+\frac{2M}{M_G}\sin^2 2\pi nt\right)\frac{d^2f}{dt^2}$$
$$=\frac{P}{M_G}\left[\cos 2\pi(N-n)t-\cos 2\pi(N+n)t\right]$$

or

$$4\pi^2\left[n_0^2-\frac{M}{M_G}n^2(1-\cos 4\pi nt)\right]f(t)+4\pi n\frac{M}{M_G}\sin 4\pi nt\,\frac{df}{dt}$$
$$+\left[1+\frac{M}{M_G}(1-\cos 4\pi nt)\right]\frac{d^2f}{dt^2}=\frac{P}{M_G}\left[\cos 2\pi(N-n)t-\cos 2\pi(N+n)t\right].$$

This is a linear differential equation in $f(t)$, but since its coefficients are not constants, a concise expression for $f(t)$ cannot be obtained as in the previous cases. A solution of this differential equation can only be given in the form of a series, the coefficients of which have to be determined by the somewhat laborious process which will now be explained and illustrated.

For the particular integral or forced oscillation let

$$f(t) = A_1 \cos 2\pi (N+n)t + A_3 \cos 2\pi (N+3n)t + A_5 \cos 2\pi (N+5n)t + \dots$$
$$+ A_{-1} \cos 2\pi (N-n)t + A_{-3} \cos 2\pi (N-3n)t + A_{-5} \cos 2\pi (N-5n)t + \dots.$$

Write this in the abbreviated form

$$f(t) = \sum_{r=-\infty}^{r=+\infty} A_r \cos 2\pi q_r t,$$

where $q_r = N + rn$, and r is any odd number, positive or negative.

By substitution it will be found that the coefficient of $\cos 2\pi q_r t$ on the left-hand side of the equation for $f(t)$ is

$$4\pi^2 \left[A_r \left\{ n_0{}^2 - \frac{M}{M_G} n^2 - \left(1 + \frac{M}{M_G}\right) q_r{}^2 \right\} + \frac{M}{2M_G} A_{r-2} q^2{}_{r-1} + \frac{M}{2M_G} A_{r+2} q^2{}_{r+1} \right]$$

or

$$\frac{M}{M_G} 2\pi^2 n_0{}^2 \left[2A_r \left\{ \frac{M_G}{M} - \left(\frac{n}{n_0}\right)^2 - \left(1 + \frac{M_G}{M}\right)\left(\frac{q_r}{n_0}\right)^2 \right\} + A_{r-2}\left(\frac{q_{r-1}}{n_0}\right)^2 + A_{r+2}\left(\frac{q_{r+1}}{n_0}\right)^2 \right].$$

Write this $\quad \dfrac{M}{M_G} 2\pi^2 n_0{}^2 [2A_r a_r + A_{r-2} b_{r-1} + A_{r+2} b_{r+1}],$

where $\qquad\qquad\qquad b_r = \left(\dfrac{q_r}{n_0}\right)^2$

and $\qquad\qquad\qquad a_r = \dfrac{M_G}{M} - \left(\dfrac{n}{n_0}\right)^2 - \left(1 + \dfrac{M_G}{M}\right) b_r.$

Equating coefficients of $\cos 2\pi q_r t$ on both sides of the equation for $f(t)$,

$$2A_r a_r + A_{r-2} b_{r-1} + A_{r+2} b_{r+1}$$

has the value $\dfrac{P}{2\pi^2 n_0{}^2 M}$, when $r = -1$, the value $-\dfrac{P}{2\pi^2 n_0{}^2 M}$, when $r = +1$, and is zero for all the other values of r.

From these conditions the coefficients in the series $\sum_{r=-\infty}^{r=+\infty} A_r \cos 2\pi q_r t$ can be determined to any desired degree of accuracy.

Thus if the series is only extended, say, from $r = -p$ to $r = +p$, the left-hand side of the equation for $f(t)$ will exceed the right by an amount

$$\frac{M}{M_G} 2\pi^2 n_0{}^2 [A_p b_{p+1} \cos 2\pi q_{p+2} t + A_{-p} b_{-(p+1)} \cos 2\pi q_{-(p+2)} t],$$

and this is equivalent to saying that an additional sinusoidal distribution of load on the girder amounting to

$$2\pi^2 n_0{}^2 \frac{M}{l}[A_p b_{p+1} \cos 2\pi q_{p+2} t + A_{-p} b_{-(p+1)} \cos 2\pi q_{-(p+2)} t] \sin \frac{\pi x}{l}$$

is required to balance the account completely.

The series for $f(t)$ must accordingly be extended in both directions until this unbalanced force is so small that its effect in producing oscillations can be ignored.

Suppose, for instance, a sufficiently good approximation for $f(t)$ is given by

$$f(t) = \sum_{r=-11}^{r=+11} A_r \cos 2\pi q_r t.$$

The equations for determining the twelve constants involved are

$$2a_1 A_1 + b_0 A_{-1} + b_2 A_3 \qquad = -\frac{P}{2\pi^2 n_0{}^2 M} \qquad \ldots\ldots(1),$$

$$2a_3 A_3 + b_2 A_1 + b_4 A_5 = 0 \qquad \ldots\ldots(2),$$

$$2a_5 A_5 + b_4 A_3 + b_6 A_7 = 0 \qquad \ldots\ldots(3),$$

$$2a_7 A_7 + b_6 A_5 + b_8 A_9 = 0 \qquad \ldots\ldots(4),$$

$$2a_9 A_9 + b_8 A_7 + b_{10} A_{11} = 0 \qquad \ldots\ldots(5),$$

$$2a_{11} A_{11} + b_{10} A_9 = 0 \qquad \ldots\ldots(6),$$

$$2a_{-1} A_{-1} + b_0 A_1 + b_{-2} A_{-3} = +\frac{P}{2\pi^2 n_0{}^2 M} \qquad \ldots\ldots(7),$$

$$2a_{-3} A_{-3} + b_{-2} A_{-1} + b_{-4} A_{-5} = 0 \qquad \ldots\ldots(8),$$

$$2a_{-5} A_{-5} + b_{-4} A_{-3} + b_{-6} A_{-7} = 0 \qquad \ldots\ldots(9),$$

$$2a_{-7} A_{-7} + b_{-6} A_{-5} + b_{-8} A_{-9} = 0 \qquad \ldots\ldots(10),$$

$$2a_{-9} A_{-9} + b_{-8} A_{-7} + b_{-10} A_{-11} = 0 \qquad \ldots\ldots(11),$$

$$2a_{-11} A_{-11} + b_{-10} A_{-9} = 0 \qquad \ldots\ldots(12).$$

From equations (6), (5), (4), (3), (2) the coefficients $A_{11}, A_9, A_7, A_5, A_3$ can be expressed in terms of A_1. From equations (12), (11), (10), (9), (8) the coefficients $A_{-11}, A_{-9}, A_{-7}, A_{-5}, A_{-3}$ can be expressed in terms of A_{-1}. Then from equations (1) and (7) the coefficients A_1 and A_{-1} can be determined and consequently all the other coefficients evaluated.

The general expressions for these coefficients are too complicated to be helpful, but their determination in any particular case presents no difficulty and this process will be illustrated by two examples in the later part of this Chapter.

The particular integral or forced oscillation obtained by the foregoing process satisfies one of the starting conditions, viz. $\frac{dy}{dt} = 0$ when $t = 0$, but it does not satisfy the other condition, that $y = 0$ when $t = 0$, and to get the true state of oscillation developed from a condition of rest and zero deflection it is necessary to superpose a complementary function or free oscillation of the type $y = f(t) \sin \frac{\pi x}{l}$, where $f(t)$ is the solution of

$$4\pi^2 \left[n_0^2 - \frac{M}{M_G} n^2 (1 - \cos 4\pi nt) \right] f(t) + 4\pi n \frac{M}{M_G} \sin 4\pi nt \frac{df}{dt}$$

$$+ \left[1 + \frac{M}{M_G} (1 - \cos 4\pi nt) \right] \frac{d^2 f}{dt^2} = 0.$$

For this solution let

$$f(t) = \sum_{r=-\infty}^{r=+\infty} A_r \cos 2\pi (N_0 + rn) t,$$

where r is any even number, positive, negative or zero, and N_0 is a number which has to be found.

Writing $N_0 + rn = q_r$, and using the notation previously adopted, on substituting

$$f(t) = \sum_{r=-\infty}^{r=+\infty} A_r \cos 2\pi q_r t$$

in the differential equation for $f(t)$, the condition that the term in $\cos 2\pi q_r t$ should be zero is

$$2a_r A_r + A_{r-2} b_{r-1} + A_{r+2} b_{r+1} = 0.$$

If a limit is put to the series in both directions, a set of equations is obtained in this way from which the ratios of the coefficients A_0, A_2, A_{-2}, A_{-4}, etc. can be determined and likewise a condition derived for finding the value of N_0.

Suppose a sufficiently accurate determination of $f(t)$ is obtained by limiting the series between the range $r = -6$ to $r = +6$. The following seven equations must then be satisfied:

$$2A_6 a_6 \ + A_4 b_5 \qquad\qquad\quad = 0 \qquad\qquad ......(1),$$
$$2A_4 a_4 \ + A_2 b_3 + A_6 b_5 \qquad = 0 \qquad\qquad ......(2),$$
$$2A_2 a_2 \ + A_0 b_1 + A_4 b_3 \qquad = 0 \qquad\qquad ......(3),$$
$$2A_{-6} a_{-6} + A_{-4} b_{-5} \qquad\qquad = 0 \qquad\qquad ......(4),$$
$$2A_{-4} a_{-4} + A_{-2} b_{-3} + A_{-6} b_{-5} = 0 \qquad\qquad ......(5),$$
$$2A_{-2} a_{-2} + A_0 b_{-1} \ + A_{-4} b_{-3} = 0 \qquad\qquad ......(6),$$
$$2A_0 a_0 \ + A_{-2} b_{-1} + A_2 b_1 \ = 0 \qquad\qquad ......(7).$$

From equations (1), (2), (3), the coefficients A_6, A_4, A_2 can be determined in terms of A_0. From equations (4), (5), (6), the coefficients A_{-6}, A_{-4}, A_{-2} can be determined in terms of A_0. Substituting in equation (7), a condition is obtained for determining $q_0 = N_0$. In performing this calculation for q_0, a "trial and error" method must be adopted, and several preliminary calculations are in general required before a value of q_0 which satisfies equation (7) is obtained. This process will be illustrated later on in this Chapter. It yields a solution which can be made to conform to any desired limits of accuracy, but it is undeniably a laborious process and an alternative method which is very much more direct and easier to carry out will now be explained. It is only an approximate method, but, when tested against the exact solution, it gives such a remarkably close approximation to the truth, that its use is justified both on the grounds of rapidity and accuracy.

APPROXIMATE METHOD FOR DETERMINING THE FREE OSCILLATIONS OF A GIRDER TRAVERSED BY A CONCENTRATED MASS MOVING UNIFORMLY ALONG THE GIRDER

This approximate method is based upon the fundamental assumption that no tractive effort has to be exerted and no work has to be expended in moving the mass along the girder at a constant speed. This assumption is not strictly correct, but the accuracy of the results when tested by the fuller analysis proves conclusively that it is a very close approximation to the truth. This assumption of no change in the total energy of the oscillating system due to the passage of the moving mass leads at once, by considerations of strain energy, to the conclusion that the amplitude of the oscillations remains unchanged by the passage of the load. The mode of vibration being always approximately a simple sine curve, the condition that the strain energy should be the same at every instant when the girder is momentarily at rest can only be satisfied if all the oscillations have the same amplitude. Again, from considerations of the kinetic energy of the system at the instants when the girder is momentarily free from deflection and strain energy, the natural frequency of the oscillations can be deduced. Suppose that when M, the concentrated mass, has moved a distance a along the girder, the frequency of the oscillations is N, and the girder is oscillating in a sine curve whose central deflection is δ. If at this instant the girder is free from deflection and strain energy, its kinetic energy is $\dfrac{1}{2}\left[\dfrac{M_G}{2} \times 4\pi^2 N^2 \delta^2\right]$. The vertical velocity of M at that instant is $2\pi N \delta \sin \dfrac{\pi a}{l}$. Hence the total kinetic energy of the oscillating system is

$$\tfrac{1}{2}Mv^2 + \tfrac{1}{2} \cdot 4\pi^2 N^2 \delta^2 \left[\frac{M_G}{2} + M \sin^2 \frac{\pi a}{l}\right].$$

Initially this kinetic energy is

$$\tfrac{1}{2}Mv^2+\tfrac{1}{2}\,.\,4\pi^2 n_0{}^2\delta^2\frac{M_G}{2}.$$

Hence, if the kinetic energy is to remain unchanged,

$$N=n_0\sqrt{\frac{M_G}{M_G+2M\sin^2\frac{\pi a}{l}}},$$

and the frequency of the oscillations depends solely upon the position of the load and not on its velocity along the girder.

This deduction, that the amplitude of the oscillations remains unchanged as M moves along the span but that the frequency varies in accordance with the relation

$$N=n_0\sqrt{\frac{M_G}{M_G+2M\sin^2\frac{\pi a}{l}}},$$

will, by a numerical example worked out later, be found to be in very close agreement with results obtained by the more exact process previously described.

NUMERICAL EXAMPLES

The foregoing series methods for determining forced and free oscillations will now be illustrated by applying them to determine the oscillations set up in a long-span bridge by the passage of a locomotive. The bridge and locomotive selected are the same as were specified in Chapter V:

Bridge data.
   Span $l=270$ feet.
   Total mass $M_G=450$ tons.
   Unloaded frequency $n_0=3$.
   Damping zero.

Locomotive data.
   Total mass $M=100$ tons.
   Circumference of driving-wheels $=15$ feet.
   Hammer-blow at N revs. per sec. $=0\cdot6N^2$ tons.

The fundamental natural frequency of the bridge when carrying 100 tons at its centre is

$$n_0\sqrt{\frac{M_G}{M_G+2M}}=3\sqrt{\frac{450}{650}}=2\cdot4961.$$

With the load placed at the quarter points on the bridge, its fundamental frequency is

$$n_0 \sqrt{\frac{M_G}{M_G + M}} = 3 \sqrt{\frac{450}{550}} = 2 \cdot 7136.$$

For any hammer-blow frequency between 3 and 2·4961, resonance will be momentarily established at some instant during the passage of the locomotive. Two cases will be studied:

Case (1) $N = 2 \cdot 4961$, giving a condition of resonance when the locomotive is at the centre of the bridge.

Case (2) $N = 2 \cdot 7136$, giving a condition of resonance when the locomotive is at the quarter points along the bridge.

Case (1)

Forced oscillation

$$N = 2 \cdot 4961, \quad v = 15 \times 2 \cdot 4961 = 37 \cdot 4415 \text{ f.s.}, \quad n = \frac{v}{2l} = 0 \cdot 0693,$$

$$P = 0 \cdot 6 \times N^2 = 3 \cdot 7383 \text{ tons}, \quad q_r = N + rn = 2 \cdot 4961 + 0 \cdot 0693r,$$

$$b_r = \left(\frac{q_r}{n_0}\right)^2 = \frac{q_r^2}{9},$$

$$a_r = \frac{M_G}{M} - \left(\frac{n}{n_0}\right)^2 - \left(1 + \frac{M_G}{M}\right) b_r = 4 \cdot 49947 - 5 \cdot 5 b_r.$$

For the forced oscillation the series $f(t) = \Sigma A_r \cos 2\pi q_r t$ will be taken to include all terms between the limits $r = -11$ and $r = +17$, a range which is sufficiently extended in both directions to give an exceedingly accurate result.

For laying down the equations by which the A coefficients are determined, the following table of values is required:

r	b_r	a_r	r	b_r	a_r
0	0·6923		0	0·6923	
1		0·4776	− 1		0·9004
2	0·7713		− 2	0·6175	
3		0·0313	− 3		1·2998
4	0·8546		− 4	0·5471	
5		− 0·4385	− 5		1·6757
6	0·9421		− 6	0·4808	
7		− 0·9340	− 7		2·0280
8	1·0339		− 8	0·4189	
9		− 1·4486	− 9		2·3576
10	1·1300		− 10	0·3612	
11		− 1·9888	− 11		2·6624
12	1·2304				
13		− 2·5525			
14	1·3350				
15		− 3·1397			
16	1·4439				
17		− 3·7504			

The equations for determining A_3, A_5, A_7, ... A_{17} in terms of A_1, and A_{-3}, A_{-5}, A_{-7}, ... A_{-11} in terms of A_{-1}, can now be laid down as follows:

$$-7{\cdot}5008A_{17} + 1{\cdot}4439A_{15} \qquad\qquad = 0 \qquad ......(1),$$
$$-6{\cdot}2794A_{15} + 1{\cdot}3350A_{13} + 1{\cdot}4439A_{17} = 0 \qquad ......(2),$$
$$-5{\cdot}1050A_{13} + 1{\cdot}2304A_{11} + 1{\cdot}3350A_{15} = 0 \qquad ......(3),$$
$$-3{\cdot}9776A_{11} + 1{\cdot}1300A_9 \; + 1{\cdot}2304A_{13} = 0 \qquad ......(4),$$
$$-2{\cdot}8972A_9 \; + 1{\cdot}0339A_7 \; + 1{\cdot}1300A_{11} = 0 \qquad ......(5),$$
$$-1{\cdot}8680A_7 \; + 0{\cdot}9421A_5 \; + 1{\cdot}0339A_9 \; = 0 \qquad ......(6),$$
$$-0{\cdot}8770A_5 \; + 0{\cdot}8546A_3 \; + 0{\cdot}9421A_7 \; = 0 \qquad ......(7),$$
$$+0{\cdot}0626A_3 \; + 0{\cdot}7713A_1 \; + 0{\cdot}8546A_5 \; = 0 \qquad ......(8),$$
$$5{\cdot}3248A_{-11} + 0{\cdot}3612A_{-9} \qquad\qquad = 0 \qquad ......(9),$$
$$4{\cdot}7152A_{-9} \; + 0{\cdot}4189A_{-7} + 0{\cdot}3612A_{-11} = 0 \qquad ......(10),$$
$$4{\cdot}0560A_{-7} \; + 0{\cdot}4808A_{-5} + 0{\cdot}4189A_{-9} \; = 0 \qquad ......(11),$$
$$3{\cdot}3514A_{-5} \; + 0{\cdot}5471A_{-3} + 0{\cdot}4808A_{-7} \; = 0 \qquad ......(12),$$
$$2{\cdot}5996A_{-3} \; + 0{\cdot}6175A_{-1} + 0{\cdot}5471A_{-5} \; = 0 \qquad ......(13).$$

The equations for determining A_1 and A_{-1} are

$$0{\cdot}9552A_1 + 0{\cdot}6923A_{-1} + 0{\cdot}7713A_3 = -\frac{P}{2\pi^2 n_0^2 M} \qquad ......(14),$$

$$1{\cdot}8008A_{-1} + 0{\cdot}6923A_1 + 0{\cdot}6175A_{-3} = +\frac{P}{2\pi^2 n_0^2 M} \qquad ......(15).$$

If D_P is the central deflection due to a steady central load P,

$$D_P = \frac{P}{2\pi^2 n_0^2 M_G},$$

and consequently $\qquad \dfrac{P}{2\pi^2 n_0^2 M} = \dfrac{M_G}{M} D_P = 4{\cdot}5 D_P.$

The values of the A coefficients in terms of D_P, obtained by solving equations (1)–(15), are as shewn in the following table:

$A_1 \; = -14{\cdot}0732D_P$	$A_{-1} = \quad 8{\cdot}6384D_P$
$A_3 \; = + \; 3{\cdot}8411D_P$	$A_{-3} = -2{\cdot}1262D_P$
$A_5 \; = +12{\cdot}4201D_P$	$A_{-5} = +0{\cdot}3532D_P$
$A_7 \; = \quad 8{\cdot}0774D_P$	$A_{-7} = -0{\cdot}0422D_P$
$A_9 \; = \quad 3{\cdot}2768D_P$	$A_{-9} = +0{\cdot}0038D_P$
$A_{11} = \quad 1{\cdot}0110D_P$	$A_{-11} = -0{\cdot}0002D_P$
$A_{13} = \quad 0{\cdot}2587D_P$	
$A_{15} = \quad 0{\cdot}0576D_P$	
$A_{17} = \quad 0{\cdot}0111D_P$	

The forced oscillation

$$y = \Sigma A_r \cos 2\pi (N + rn) t \sin \frac{\pi x}{l}$$

can be written in the form

$$y = [\cos 2\pi Nt \, \Sigma \, (A_r + A_{-r}) \cos 2\pi rnt + \sin 2\pi Nt \, \Sigma \, (A_{-r} - A_r) \sin 2\pi rnt] \sin \frac{\pi x}{l},$$

and for the case under consideration this becomes

$$y = D_p \cos 2\pi Nt \, [-5 \cdot 4348 \cos 2\pi nt + 1 \cdot 7149 \cos 6\pi nt + 12 \cdot 7733 \cos 10\pi nt$$

$$+ 8 \cdot 0352 \cos 14\pi nt + 3 \cdot 2806 \cos 18\pi nt + 1 \cdot 0108 \cos 22\pi nt$$

$$+ 0 \cdot 2587 \cos 26\pi nt + 0 \cdot 0576 \cos 30\pi nt + 0 \cdot 0111 \cos 34\pi nt] \sin \frac{\pi x}{l}$$

$$+ D_p \sin 2\pi Nt \, [22 \cdot 7116 \sin 2\pi nt - 5 \cdot 9673 \sin 6\pi nt - 12 \cdot 0669 \sin 10\pi nt$$

$$- 8 \cdot 1196 \sin 14\pi nt - 3 \cdot 2730 \sin 18\pi nt - 1 \cdot 0112 \sin 22\pi nt$$

$$- 0 \cdot 2587 \sin 26\pi nt - 0 \cdot 0576 \sin 30\pi nt - 0 \cdot 0111 \sin 34\pi nt] \sin \frac{\pi x}{l}.$$

Let δ denote the central deflection when the load has advanced a distance a along the bridge. Since $2\pi nt = \dfrac{\pi vt}{l} = \dfrac{\pi a}{l}$, the connection between δ and a is given by

$$\delta = D_p \cos 2\pi Nt \left[-5 \cdot 43 \cos \frac{\pi a}{l} + 1 \cdot 71 \cos \frac{3\pi a}{l} + 12 \cdot 77 \cos \frac{5\pi a}{l} \right.$$

$$+ 8 \cdot 04 \cos \frac{7\pi a}{l} + 3 \cdot 28 \cos \frac{9\pi a}{l} + 1 \cdot 01 \cos \frac{11\pi a}{l} + 0 \cdot 26 \cos \frac{13\pi a}{l}$$

$$\left. + 0 \cdot 06 \cos \frac{15\pi a}{l} + 0 \cdot 01 \cos \frac{17\pi a}{l} \right]$$

$$+ D_p \sin 2\pi Nt \left[22 \cdot 71 \sin \frac{\pi a}{l} - 5 \cdot 97 \sin \frac{3\pi a}{l} - 12 \cdot 07 \sin \frac{5\pi a}{l} \right.$$

$$- 8 \cdot 12 \sin \frac{7\pi a}{l} - 3 \cdot 27 \sin \frac{9\pi a}{l} - 1 \cdot 01 \sin \frac{11\pi a}{l} - 0 \cdot 26 \sin \frac{13\pi a}{l}$$

$$\left. - 0 \cdot 06 \sin \frac{15\pi a}{l} - 0 \cdot 01 \sin \frac{17\pi a}{l} \right].$$

Now D_p is the central deflection due to a steady central load of $3 \cdot 7383$ tons. If D_W denotes the central deflection due to the weight of the moving mass of 100 tons, when standing at the centre of the bridge, $D_p = \dfrac{3 \cdot 7383}{100} D_W$.

Expressing δ in the form

$$\delta = \frac{D_W}{100} [\Phi(a) \cos 2\pi Nt + \Psi(a) \sin 2\pi Nt],$$

where $N = 2\cdot4961$, the values of $\Phi(a)$ and $\Psi(a)$ are as tabulated below:

a	0	$\dfrac{l}{16}$	$\dfrac{l}{8}$	$\dfrac{3l}{16}$	$\dfrac{l}{4}$	$\dfrac{5l}{16}$	$\dfrac{3l}{8}$	$\dfrac{7l}{16}$	$\dfrac{l}{2}$
$\Phi(a)$	81·16	12·22	−74·53	−71·17	−25·88	6·66	16·93	12·08	0
$\Psi(a)$	0	−78·80	−32·28	50·07	87·11	90·20	85·88	83·68	83·21

a	l	$\dfrac{15l}{16}$	$\dfrac{7l}{8}$	$\dfrac{13l}{16}$	$\dfrac{3l}{4}$	$\dfrac{11l}{16}$	$\dfrac{5l}{8}$	$\dfrac{9l}{16}$	$\dfrac{l}{2}$
$\Phi(a)$	−81·16	−12·22	74·53	71·17	25·88	−6·66	−16·93	−12·08	0
$\Psi(a)$	0	−78·80	−32·28	50·07	87·11	90·20	85·88	83·68	83·21

In Fig. 16 (a) the curves $\delta = \dfrac{D_W}{100}\Phi(a)\cos 2\pi Nt$ and $\delta = \dfrac{D_W}{100}\Psi(a)\sin 2\pi Nt$

are plotted and the superposition of these gives the forced oscillation shewn by the full-line curve of Fig. 16 (b).

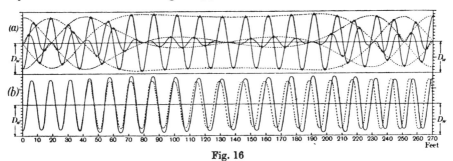

Fig. 16

Before proceeding to calculate the free oscillation for Case (1), the forced oscillation for Case (2) will be determined.

Case (2)

Forced oscillation

$$N = 2\cdot7136, \quad v = 15 \times 2\cdot7136 = 40\cdot7040 \text{ f.s.}, \quad n = \frac{v}{2l} = 0\cdot0754,$$

$$P = 0\cdot6 N^2 = 4\cdot4182 \text{ tons},$$

$$q_r = N + rn = 2\cdot7136 + 0\cdot0754 r,$$

$$b_r = \left(\frac{q_r}{n_0}\right)^2 = \frac{q_r^2}{9},$$

$$a_r = \frac{M_G}{M} - \left(\frac{n}{n_0}\right)^2 - \left(1 + \frac{M_G}{M}\right) b_r = 4\cdot4994 - 5\cdot5 b_r.$$

In this case a sufficiently accurate result will be obtained if the series $f(t) = \Sigma A_r \cos 2\pi q_r t$ is limited within the range $r = -11$ to $r = +11$.

For laying down the equations by which the A coefficients are determined, the following table of the values of a_r and b_r is required:

r	b_r	a_r	r	b_r	a_r
0	0·8182		0	0·8182	
1		−0·2543	−1		0·2462
2	0·9116		−2	0·7298	
3		−0·7823	−3		0·7182
4	1·0102		−4	0·6464	
5		−1·3378	−5		1·1631
6	1·1137		−6	0·5681	
7		−1·9213	−7		1·5794
8	1·2224		−8	0·4949	
9		−2·5329	−9		1·9688
10	1·3360		−10	0·4267	
11		−3·1720	−11		2·3296

The equations for determining the A coefficients are as follows:

$$-6·3440A_{11} + 1·3360A_9 \qquad\qquad = 0 \qquad\qquad ......(1),$$
$$-5·0658A_9 + 1·2224A_7 + 1·3660A_{11} = 0 \qquad\qquad ......(2),$$
$$-3·8426A_7 + 1·1137A_5 + 1·2224A_9 = 0 \qquad\qquad ......(3),$$
$$-2·6756A_5 + 1·0102A_3 + 1·1137A_7 = 0 \qquad\qquad ......(4),$$
$$-1·5646A_3 + 0·9116A_1 + 1·0102A_5 = 0 \qquad\qquad ......(5),$$
$$-0·5086A_1 + 0·8182A_{-1} + 0·9116A_3 = -4·5D_P \qquad ......(6),$$
$$4·6592A_{-11} + 0·4267A_{-9} \qquad\qquad = 0 \qquad\qquad ......(7),$$
$$3·9376A_{-9} + 0·4949A_{-7} + 0·4267A_{-11} = 0 \qquad\qquad ......(8),$$
$$3·1588A_{-7} + 0·5681A_{-5} + 0·4949A_{-9} = 0 \qquad\qquad ......(9),$$
$$2·3262A_{-5} + 0·6464A_{-3} + 0·5681A_{-7} = 0 \qquad\qquad ......(10),$$
$$1·4364A_{-3} + 0·7298A_{-1} + 0·6464A_{-5} = 0 \qquad\qquad ......(11),$$
$$0·4924A_{-1} + 0·8182A_1 + 0·7298A_{-3} = 4·5D_P \qquad\qquad ......(12),$$

where D_P is the central deflection due to a steady central load $P = 4·4182$ tons.

The values of the A coefficients obtained by solving these equations (1)–(12) are shewn in the following table:

$A_1 = 6·0666D_P$	$A_{-1} = -7·2071D_P$
$A_3 = 4·9150D_P$	$A_{-3} = +4·2234D_P$
$A_5 = 2·1361D_P$	$A_{-5} = -1·2492D_P$
$A_7 = 0·6738D_P$	$A_{-7} = +0·3090D_P$
$A_9 = 0·1722D_P$	$A_{-9} = -0·0392D_P$
$A_{11} = 0·0363D_P$	$A_{-11} = +0·0036D_P$

The forced oscillation

$$y = \Sigma A_r \cos 2\pi (N + rn)\, t \sin \frac{\pi x}{l}$$

can be written in the form

$$y = [\cos 2\pi Nt \, \Sigma (A_r + A_{-r}) \cos 2\pi rnt + \sin 2\pi Nt \, \Sigma (A_{-r} - A_r) \sin 2\pi rnt] \sin \frac{\pi x}{l},$$

and for the case under consideration this becomes

$$y = D_P \cos 2\pi Nt \,[- 1 \cdot 1405 \cos 2\pi nt + 9 \cdot 1384 \cos 6\pi nt + 0 \cdot 8869 \cos 10\pi nt$$

$$+ 0 \cdot 9828 \cos 14\pi nt + 0 \cdot 1330 \cos 18\pi nt + 0 \cdot 0399 \cos 22\pi nt] \sin \frac{\pi x}{l}$$

$$- D_P \sin 2\pi Nt \,[13 \cdot 2737 \sin 2\pi nt + 0 \cdot 6916 \sin 6\pi nt + 3 \cdot 3853 \sin 10\pi nt$$

$$+ 0 \cdot 3648 \sin 14\pi nt + 0 \cdot 2114 \sin 18\pi nt + 0 \cdot 0327 \sin 22\pi nt] \sin \frac{\pi x}{l}.$$

Let δ denote the central deflection when the load has advanced a distance a along the bridge. Since $2\pi nt = \dfrac{\pi vt}{l} = \dfrac{\pi a}{l}$, the connection between δ and a is given by

$$\delta = D_P \cos 2\pi Nt \left[- 1 \cdot 14 \cos \frac{\pi a}{l} + 9 \cdot 14 \cos \frac{3\pi a}{l} + 0 \cdot 89 \cos \frac{5\pi a}{l} \right.$$

$$\left. + 0 \cdot 98 \cos \frac{7\pi a}{l} + 0 \cdot 13 \cos \frac{9\pi a}{l} + 0 \cdot 04 \cos \frac{11\pi a}{l} \right]$$

$$- D_P \sin 2\pi Nt \left[13 \cdot 27 \sin \frac{\pi a}{l} + 0 \cdot 69 \sin \frac{3\pi a}{l} + 3 \cdot 39 \sin \frac{5\pi a}{l} \right.$$

$$\left. + 0 \cdot 36 \sin \frac{7\pi a}{l} + 0 \cdot 21 \sin \frac{9\pi a}{l} + 0 \cdot 03 \sin \frac{11\pi a}{l} \right].$$

Now D_P is the central deflection due to a steady central load of $4 \cdot 4182$ tons. If D_W denotes the central deflection due to the weight of the moving mass of 100 tons when standing at the centre of the bridge, $D_P = \dfrac{4 \cdot 4182}{100} D_W$.

Expressing δ in the form

$$\delta = \frac{D_W}{100} [\Phi (a) \cos 2\pi Nt + \Psi (a) \sin 2\pi Nt],$$

where $N = 2 \cdot 7136$, the values of $\Phi (a)$ and $\Psi (a)$ in terms of a are as tabulated below:

a	0	$\dfrac{l}{16}$	$\dfrac{l}{8}$	$\dfrac{3l}{16}$	$\dfrac{l}{4}$	$\dfrac{5l}{16}$	$\dfrac{3l}{8}$	$\dfrac{7l}{16}$	$\dfrac{l}{2}$
$\Phi (a)$	$44 \cdot 36$	$31 \cdot 57$	$4 \cdot 71$	$-17 \cdot 83$	$-31 \cdot 55$	$-38 \cdot 55$	$-37 \cdot 32$	$-23 \cdot 95$	0
$\Psi (a)$	0	$-28 \cdot 13$	$-39 \cdot 20$	$-36 \cdot 37$	$-32 \cdot 67$	$-35 \cdot 99$	$-47 \cdot 95$	$-62 \cdot 85$	$-69 \cdot 72$

a	l	$\dfrac{15l}{16}$	$\dfrac{7l}{8}$	$\dfrac{13l}{16}$	$\dfrac{3l}{4}$	$\dfrac{11l}{16}$	$\dfrac{5l}{8}$	$\dfrac{9l}{16}$	$\dfrac{l}{2}$
$\Phi (a)$	$-44 \cdot 36$	$-31 \cdot 57$	$-4 \cdot 71$	$17 \cdot 83$	$31 \cdot 55$	$38 \cdot 55$	$37 \cdot 32$	$23 \cdot 95$	0
$\Psi (a)$	0	$-28 \cdot 13$	$-39 \cdot 20$	$-36 \cdot 37$	$-32 \cdot 67$	$-35 \cdot 99$	$-47 \cdot 95$	$-62 \cdot 85$	$-69 \cdot 72$

The curves $\delta = \dfrac{D_W}{100}\,\Phi\,(a)\cos 2\pi Nt$ and $\delta = \dfrac{D_W}{100}\,\Psi\,(a)\sin 2\pi Nt$ are plotted in Fig. 17 (a) and the result of superposing them to give the forced oscillation is shewn by the full-line curve on Fig. 17 (b). It will be seen that the amplitude of the oscillations is considerably reduced in comparison with Case (1), Fig. 16, in which synchronism occurs when the load reaches the centre of the bridge.

The particular integrals or forced oscillations, determined for Cases (1) and (2), satisfy the condition that $\dfrac{dy}{dt} = 0$ when $t = 0$, but to adjust the other starting condition, $y = 0$ when $t = 0$, a complementary function or oscillation

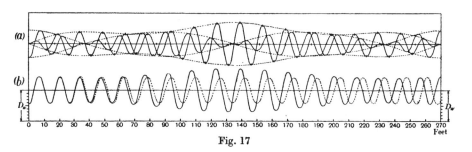

Fig. 17

corresponding to the case of zero hammer-blow has to be superposed, and this, although it includes the inertia effect of the moving load, will be termed a free oscillation. The method described earlier in this Chapter will now be applied to determine the free oscillations corresponding to Cases (1) and (2).

Case (1)

Free oscillation

$$v = 37\cdot4415 \text{ f.s.}, \quad n = \frac{v}{2l} = 0\cdot0693.$$

The state of free oscillation can be expressed in the form

$$y = \Sigma A_r \cos 2\pi\,(N_0 + rn)\,t \sin\frac{\pi x}{l},$$

where r is any even number, positive, negative or zero, and N_0 is a number which has to be found by a trial and error process. For this particular case a preliminary investigation shewed that N_0 was very nearly $2\cdot5920$. The limits selected for the series, as giving an exceedingly accurate representation of the free oscillation, are $r = -6$ to $r = +12$.

Writing

$$q_r = N_0 + rn, \quad b_r = \left(\frac{q_r}{n_0}\right)^2 \text{ and } a_r = \frac{M_G}{M} - \left(\frac{n}{n_0}\right)^2 - \left(1 + \frac{M}{M_G}\right)b_r,$$

the values for a_r and b_r for this particular case $N_0 = 2\cdot5920$ are given in the following table:

r	b_r	a_r	r	b_r	a_r
0		0·3937	0		0·3937
1	0·7869		−1	0·7071	
2		−0·0571	−2		0·8211
3	0·8710		−3	0·6315	
4		−0·5314	−4		1·2250
5	0·9594		−5	0·5602	
6		−1·0292	−6		1·6053
7	1·0521				
8		−1·5504			
9	1·1490				
10		−2·0952			
11	1·2501				
12		−2·6634			

The equations for determining $A_{12}, A_{10}, \ldots A_2$ and A_{-6}, A_{-4}, A_{-2} in terms of A_0 are:

$$-5\cdot3268A_{12} + 1\cdot2501A_{10} \qquad\qquad = 0,$$
$$-4\cdot1904A_{10} + 1\cdot1490A_8 \;+ 1\cdot2501A_{12} = 0,$$
$$-3\cdot1008A_8 \;+ 1\cdot0521A_6 \;+ 1\cdot1490A_{10} = 0,$$
$$-2\cdot0584A_6 \;+ 0\cdot9594A_4 \;+ 1\cdot0521A_8 = 0,$$
$$-1\cdot0628A_4 \;+ 0\cdot8710A_2 \;+ 0\cdot9594A_6 = 0,$$
$$-0\cdot1142A_2 \;+ 0\cdot7869A_0 \;+ 0\cdot8710A_4 = 0,$$

$$3\cdot2106A_{-6} + 0\cdot5602A_{-4} \qquad\qquad = 0,$$
$$2\cdot4500A_{-4} + 0\cdot6315A_{-2} + 0\cdot5602A_{-6} = 0,$$
$$1\cdot6422A_{-2} + 0\cdot7071A_0 \;+ 0\cdot6315A_{-4} = 0.$$

From these equations $A_{12}, A_{10}, \ldots A_2$ and A_{-6}, A_{-4}, A_{-2} can be expressed in terms of A_0, and the remaining equation, $2A_0a_0 + A_2b_1 + A_{-2}b_{-1} = 0$, is the test for determining if $N_0 = 2\cdot5920$ is a correct assumption. In this case

$$2A_0a_0 + A_2b_1 + A_{-2}b_{-1} = A_0[0\cdot7874 - 0\cdot4485 - 0\cdot3395] = -A_0[0\cdot0006].$$

Hence the assumption that $N_0 = 2\cdot5920$ is exceedingly near the truth.

The values of the coefficients in terms of A_0 are as follows:

$A_2 = -0\cdot5700A_0$	$A_{-2} = -0\cdot4801A_0$
$A_4 = -0\cdot9782A_0$	$A_{-4} = +0\cdot1289A_0$
$A_6 = -0\cdot5662A_0$	$A_{-6} = -0\cdot0225A_0$
$A_8 = -0\cdot2157A_0$	
$A_{10} = -0\cdot0636A_0$	
$A_{12} = -0\cdot0149A_0$	

Employing the transformation

$$\Sigma A_r \cos 2\pi (N_0 + rn)t$$
$$= \cos 2\pi N_0 t \, \Sigma (A_r + A_{-r}) \cos 2\pi rnt + \sin 2\pi N_0 t \, \Sigma (A_{-r} - A_r) \sin 2\pi rnt,$$

the equation for the free oscillation can be written in the form

$$y = A_0 \cos 2\pi N_0 t \, [1 - 1{\cdot}0501 \cos 4\pi nt - 0{\cdot}8493 \cos 8\pi nt - 0{\cdot}5887 \cos 12\pi nt$$
$$- 0{\cdot}2157 \cos 16\pi nt - 0{\cdot}0636 \cos 20\pi nt - 0{\cdot}0149 \cos 24\pi nt] \sin \frac{\pi x}{l}$$
$$+ A_0 \sin 2\pi N_0 t \, [0{\cdot}0899 \sin 4\pi nt + 1{\cdot}1071 \sin 8\pi nt + 0{\cdot}5437 \sin 12\pi nt$$
$$+ 0{\cdot}2157 \sin 16\pi nt + 0{\cdot}0636 \sin 20\pi nt + 0{\cdot}0149 \sin 24\pi nt] \sin \frac{\pi x}{l}.$$

The connection between the central deflection δ and a, the distance the load has advanced along the bridge, is accordingly given by

$$\delta = A_0 \cos 2\pi N_0 t \left[1 - 1{\cdot}05 \cos \frac{2\pi a}{l} - 0{\cdot}85 \cos \frac{4\pi a}{l} - 0{\cdot}59 \cos \frac{6\pi a}{l} \right.$$
$$\left. - 0{\cdot}22 \cos \frac{8\pi a}{l} - 0{\cdot}06 \cos \frac{10\pi a}{l} - 0{\cdot}01 \cos \frac{12\pi a}{l} \right]$$
$$+ A_0 \sin 2\pi N_0 t \left[0{\cdot}09 \sin \frac{2\pi a}{l} + 1{\cdot}11 \sin \frac{4\pi a}{l} + 0{\cdot}54 \sin \frac{6\pi a}{l} \right.$$
$$\left. + 0{\cdot}22 \sin \frac{8\pi a}{l} + 0{\cdot}06 \sin \frac{10\pi a}{l} + 0{\cdot}01 \sin \frac{12\pi a}{l} \right].$$

Expressing δ in the form

$$A_0 [\Phi (a) \cos 2\pi N_0 t + \Psi (a) \sin 2\pi N_0 t],$$

where $N_0 = 2{\cdot}5920$, the values of $\Phi (a)$ and $\Psi (a)$ in terms of a are as tabulated below:

a	0	$\frac{l}{16}$	$\frac{l}{8}$	$\frac{3l}{16}$	$\frac{l}{4}$	$\frac{5l}{16}$	$\frac{3l}{8}$	$\frac{7l}{16}$	$\frac{l}{2}$
$\Phi (a)$	$-1{\cdot}78$	$-0{\cdot}77$	$0{\cdot}94$	$1{\cdot}68$	$1{\cdot}64$	$1{\cdot}51$	$1{\cdot}50$	$1{\cdot}58$	$1{\cdot}62$
$\Psi (a)$	0	$1{\cdot}60$	$1{\cdot}50$	$0{\cdot}43$	$-0{\cdot}39$	$-0{\cdot}72$	$-0{\cdot}70$	$-0{\cdot}42$	0
a	l	$\frac{15l}{16}$	$\frac{7l}{8}$	$\frac{13l}{16}$	$\frac{3l}{4}$	$\frac{11l}{16}$	$\frac{5l}{8}$	$\frac{9l}{16}$	$\frac{l}{2}$
$\Phi (a)$	$-1{\cdot}78$	$-0{\cdot}77$	$0{\cdot}94$	$1{\cdot}68$	$1{\cdot}64$	$1{\cdot}51$	$1{\cdot}50$	$1{\cdot}58$	$1{\cdot}62$
$\Psi (a)$	0	$-1{\cdot}60$	$-1{\cdot}50$	$-0{\cdot}43$	$0{\cdot}39$	$0{\cdot}72$	$0{\cdot}70$	$0{\cdot}42$	0

The component curves $\delta = \Phi (a) \cos 2\pi N_0 t$ and $\delta = \Psi (a) \sin 2\pi N_0 t$ are plotted in Fig. 18(a) and superposed to give the free oscillation shewn in Fig. 18(b). The frequency thus obtained agrees so closely with the formula

$$N = n_0 \sqrt{\frac{M_G}{M_G + 2M \sin^2 \frac{\pi a}{l}}},$$

which was obtained by the approximation method, that, within the errors of draughtsmanship, no discrepancy can be detected. The oscillations shew a small variation in amplitude, but it only amounts to a departure of about 5 per cent. from the mean, and, for all practical purposes, a sufficiently good approximation to the state of free oscillation is attained by taking the

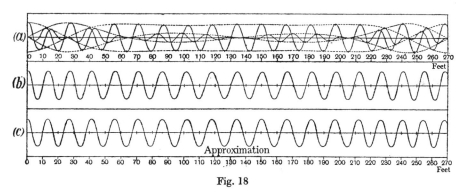

Fig. 18

oscillations to be of constant magnitude and their frequency N to vary in accordance with the formula

$$N = n_0 \sqrt{\frac{M_G}{M_G + 2M \sin^2 \frac{\pi a}{l}}}.$$

The free oscillation as obtained by the approximate process is shewn by Fig. 18(c) and the difference between that and Fig. 18(b) is barely distinguishable.

Case (2)

Free oscillation

$$v = 40 \cdot 7043, \quad n = \frac{v}{2l} = 0 \cdot 0754.$$

Expressing the state of oscillation in the form

$$y = \Sigma A_r \cos 2\pi (N_0 + rn) t \sin \frac{\pi x}{l},$$

where r is any even number, positive, negative or zero, a preliminary process of trial and error shewed that N_0 was very nearly $2 \cdot 580$. The limits found necessary to give an accurate representation of the oscillation are $r = -6$ to $r = +12$. The values of a_r and b_r corresponding to $N_0 = 2\cdot580$ are shewn in the following table:

r	b_r	a_r	r	b_r	a_r
0		0·4317	0		0·4317
1	0·7835		-1	0·6970	
2		$-0\cdot0577$	-2		0·8933
3	0·8750		-3	0·6156	
4		$-0\cdot5750$	-4		1·3271
5	0·9715		-5	0·5329	
6		$-1\cdot1200$	-6		1·7332
7	1·0732				
8		$-1\cdot6928$			
9	1·1798				
10		$-2\cdot2934$			
11	1·2916				
12		$-2\cdot9218$			

The equations for determining $A_{12}, A_{10}, \ldots A_2$ and A_{-6}, A_{-4}, A_{-2} in terms of A_0 are:

$$-5\cdot8436A_{12} + 1\cdot2916A_{10} = 0,$$
$$-4\cdot5868A_{10} + 1\cdot1798A_8 + 1\cdot2916A_{12} = 0,$$
$$-3\cdot3856A_8 + 1\cdot0732A_6 + 1\cdot1798A_{10} = 0,$$
$$-2\cdot2400A_6 + 0\cdot9715A_4 + 1\cdot0732A_8 = 0,$$
$$-1\cdot1500A_4 + 0\cdot8750A_2 + 0\cdot9715A_6 = 0,$$
$$-0\cdot1154A_2 + 0\cdot7835A_0 + 0\cdot8750A_4 = 0,$$
$$3\cdot4664A_{-6} + 0\cdot5329A_{-4} = 0,$$
$$2\cdot6542A_{-4} + 0\cdot6156A_{-2} + 0\cdot5329A_{-6} = 0,$$
$$1\cdot7866A_{-2} + 0\cdot6970A_0 + 0\cdot6156A_{-4} = 0.$$

From these nine equations the following results are obtained:

$$A_{12} = 0\cdot2210A_{10}, \quad A_{10} = 0\cdot2743A_8, \quad A_8 = 0\cdot3505A_6,$$
$$A_6 = 0\cdot5212A_4, \quad A_4 = 1\cdot3595A_2, \quad A_2 = -0\cdot7294A_0,$$

and $\quad A_{-6} = -0\cdot1537A_{-4}, \quad A_{-4} = -0\cdot2393A_{-2}, \quad A_{-2} = -0\cdot4252A_0.$

The remaining equation, $2A_0a_0 + A_2b_1 + A_{-2}b_{-1} = 0$, is the test for determining if the assumption $N_0 = 2\cdot580$ is correct. In this case

$$2A_0a_0 + A_2b_1 + A_{-2}b_{-1} = A_0[0\cdot8634 - 0\cdot29636 - 0\cdot51748] = -A_0[0\cdot0044],$$

and the assumption satisfies the test in a satisfactory manner.

The values of the coefficients in terms of A_0 are as follows:

$A_2 = -0\cdot7294A_0$	$A_{-2} = -0\cdot4252A_0$
$A_4 = -0\cdot9916A_0$	$A_{-4} = +0\cdot1018A_0$
$A_6 = -0\cdot5168A_0$	$A_{-6} = -0\cdot0156A_0$
$A_8 = -0\cdot1811A_0$	
$A_{10} = -0\cdot0497A_0$	
$A_{12} = -0\cdot0110A_0$	

Expressing the oscillation in the form

$$y = \cos 2\pi N_0 t \left[\Sigma \left(A_r + A_{-r} \right) \cos 2\pi rnt \right] \sin \frac{\pi x}{l}$$

$$+ \sin 2\pi N_0 t \left[\Sigma \left(A_{-r} - A_r \right) \sin 2\pi rnt \right] \sin \frac{\pi x}{l},$$

for this particular case

$$y = A_0 \cos 2\pi N_0 t \left[1 - 1 \cdot 1546 \cos 4\pi nt - 0 \cdot 8898 \cos 8\pi nt - 0 \cdot 5324 \cos 12\pi nt \right.$$

$$\left. - 0 \cdot 1811 \cos 16\pi nt - 0 \cdot 0497 \cos 20\pi nt - 0 \cdot 0110 \cos 24\pi nt \right] \sin \frac{\pi x}{l}$$

$$+ A_0 \sin 2\pi N_0 t \left[0 \cdot 3042 \sin 4\pi nt + 1 \cdot 0934 \sin 8\pi nt + 0 \cdot 5012 \sin 12\pi nt \right.$$

$$\left. + 0 \cdot 1811 \sin 16\pi nt + 0 \cdot 0497 \sin 20\pi nt + 0 \cdot 0110 \sin 24\pi nt \right] \sin \frac{\pi x}{l}.$$

The connection between the central deflection δ and a, the distance the load has advanced along the bridge, is accordingly given by

$$\delta = A_0 \cos 2\pi N_0 t \left[1 - 1 \cdot 15 \cos \frac{2\pi a}{l} - 0 \cdot 89 \cos \frac{4\pi a}{l} - 0 \cdot 53 \cos \frac{6\pi a}{l} \right.$$

$$\left. - 0 \cdot 18 \cos \frac{8\pi a}{l} - 0 \cdot 05 \cos \frac{10\pi a}{l} - 0 \cdot 01 \cos \frac{12\pi a}{l} \right]$$

$$+ A_0 \sin 2\pi N_0 t \left[0 \cdot 30 \sin \frac{2\pi a}{l} + 1 \cdot 09 \sin \frac{4\pi a}{l} + 0 \cdot 50 \sin \frac{6\pi a}{l} \right.$$

$$\left. + 0 \cdot 18 \sin \frac{8\pi a}{l} + 0 \cdot 05 \sin \frac{10\pi a}{l} + 0 \cdot 01 \sin \frac{12\pi a}{l} \right].$$

Expressing δ in the form

$$\delta = A_0 \left[\Phi \left(a \right) \cos 2\pi N_0 t + \Psi \left(a \right) \sin 2\pi N_0 t \right],$$

where $N_0 = 2 \cdot 580$, the values of $\Phi \left(a \right)$ and $\Psi \left(a \right)$ in terms of a are as tabulated below:

a	0	$\dfrac{l}{16}$	$\dfrac{l}{8}$	$\dfrac{3l}{16}$	$\dfrac{l}{4}$	$\dfrac{5l}{16}$	$\dfrac{3l}{8}$	$\dfrac{7l}{16}$	$\dfrac{l}{2}$
$\Phi \left(a \right)$	$-1 \cdot 81$	$-0 \cdot 87$	$0 \cdot 78$	$1 \cdot 62$	$1 \cdot 72$	$1 \cdot 62$	$1 \cdot 58$	$1 \cdot 62$	$1 \cdot 65$
$\Psi \left(a \right)$	0	$1 \cdot 58$	$1 \cdot 61$	$0 \cdot 66$	$-0 \cdot 15$	$-0 \cdot 53$	$-0 \cdot 54$	$-0 \cdot 33$	0

a	l	$\dfrac{15l}{16}$	$\dfrac{7l}{8}$	$\dfrac{13l}{16}$	$\dfrac{3l}{4}$	$\dfrac{11l}{16}$	$\dfrac{5l}{8}$	$\dfrac{9l}{16}$	$\dfrac{l}{2}$
$\Phi \left(a \right)$	$-1 \cdot 81$	$-0 \cdot 87$	$0 \cdot 78$	$1 \cdot 62$	$1 \cdot 72$	$1 \cdot 62$	$1 \cdot 58$	$1 \cdot 62$	$1 \cdot 65$
$\Psi \left(a \right)$	0	$-1 \cdot 58$	$-1 \cdot 61$	$-0 \cdot 66$	$0 \cdot 15$	$0 \cdot 53$	$0 \cdot 54$	$0 \cdot 33$	0

The result of plotting the component curves $\delta = \Phi \left(a \right) \cos 2\pi N_0 t$ and $\delta = \Psi \left(a \right) \sin 2\pi N_0 t$, and then superposing them to give the free oscillation, is shewn in Figs. 19(a) and 19(b). The free oscillation, as obtained by assuming a constant amplitude and a frequency given by

$$N = n_0 \sqrt{\dfrac{M_G}{M_G + 2M \sin^2 \dfrac{\pi a}{l}}},$$

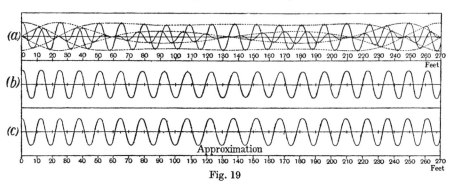

Fig. 19

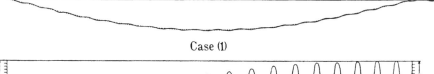

Case (1)

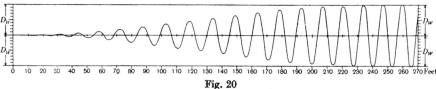

Fig. 20

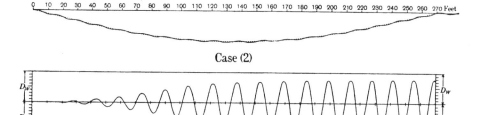

Case (2)

Fig. 21

is shewn by Fig. 19(c), and, as in Case (1), the agreement with the exact method leaves little to be desired.

The deflection produced by the steady moving load of 100 tons is obtained by superposing on the "crawl-deflection" a state of free oscillation to satisfy the starting conditions. The results thus obtained, for Cases (1) and (2), are shewn by the upper curves of Figs. 20 and 21. The lower curves on Figs. 20

and 21 shew how, for Cases (1) and (2), the oscillations due to hammer-blow mount up from an initial condition of rest and zero deflection. These latter curves are obtained by combining the forced oscillations deduced in Figs. 16 and 17 with the free oscillations shewn dotted in conjunction with these forced oscillations. The state of oscillation due to the combined effect of hammer-blow and the gravity force of the moving load is given for Case (1) by Fig. 22 and for Case (2) by Fig. 23—Fig. 22 being the superposition of the two curves of Fig. 20 and Fig. 23 being the superposition of the two curves of Fig. 21. As might be anticipated, the oscillations for Case (2) mount up more rapidly at the start than those for Case (1), but eventually the latter predominate. The maximum deflection in the two cases is, however, practically identical and has the value of $1 \cdot 75 D_W$.

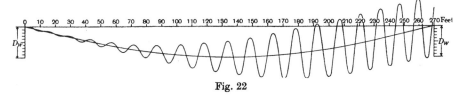

Fig. 22

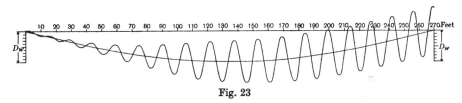

Fig. 23

Comparison should here be made with Fig. 12, Chapter V, which represents the most violent state of oscillation developed in this same bridge if the inertia effect of the moving load is ignored. It will be seen by this comparison that inertia reduces the dynamic effect to a very marked extent. This is partly due to the fact that the inertia of the moving load reduces the speed at which resonance is developed, but the reduction is mainly caused by the variation in the natural frequency of the loaded bridge, in consequence of which a true condition of resonance can only be momentarily attained.

The validity of the mathematical analysis contained in this Chapter was tested by experiments on a model bridge at the Cambridge Engineering Department. Fig. 24 is a diagrammatic illustration of the apparatus. The bridge took the form of two parallel rectangular steel bars giving a span of 12 feet and, to minimize damping, the bars rested on knife edges at the two ends. The moving load was a springless two wheeled truck stabilized by a small trailing wheel which ran on a centre rail supported independently of the bridge. The hammer-blow effect was produced by motor-driven spur-wheels rotating in opposite directions and carrying balance-weights. The frequency and magnitude of the hammer-blows could accordingly be altered at will independently of the speed of the truck. The truck in its approach run was accelerated by a falling weight, but it crossed the bridge at a constant speed. The unloaded fundamental frequency of the

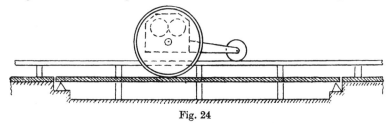

Fig. 24

bridge was 5·7 periods per sec.; the ratio of the mass of the truck to the mass of the bridge was 0·4, and these characteristics correspond approximately to those of a railway bridge having a span of about 120 feet. The variation of the central deflection as the truck crossed the bridge was recorded by a needle on a moving celluloid film, and the records thus obtained were found to be in excellent agreement with those predicted by the theory in this Chapter, for all ranges of speed.

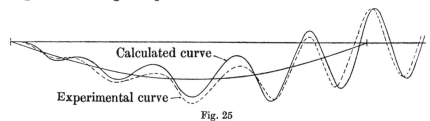

Fig. 25

Fig. 25 illustrates the nature of the agreement attained, the full-line curve being the deflection deduced by analysis, and the dotted curve being that obtained by experiment. The agreement in this particular case (which is typical of numerous records) is as good as could be expected, and the small lag developed by the experimental curve for the middle part of the run is probably due to the fact that while damping was neglected in the calculation, in practice it was not entirely eliminated.

CHAPTER VIII

OSCILLATIONS PRODUCED BY A MOVING ALTERNATING FORCE ASSOCIATED WITH A MOVING CONCENTRATED MASS AND DAMPING INFLUENCES

Under this heading is included all the characteristics necessary to take into account when predicting the oscillations produced in a long-span bridge, by the passage of a locomotive, assuming that the oscillations are not sufficiently active to overcome the friction in the spring movement of the locomotive. The problem to be investigated is illustrated by Fig. 15 of Chapter VII, and following the notation previously adopted, the primary and dominating component of the state of oscillation is given by the equation

$$EI\frac{d^4y}{dx^4} + 4\pi n_b m\frac{dy}{dt} + m\frac{d^2y}{dt^2} = \frac{2}{l}[P\sin 2\pi Nt - M\alpha]\sin 2\pi nt \sin\frac{\pi x}{l},$$

where α is the downward vertical acceleration of the moving mass M.

Assuming a solution of the form $y = f(t)\sin\frac{\pi x}{l}$, α is given by

$$\alpha = -4\pi^2 n^2 \sin 2\pi nt\, f(t) + 4\pi n \cos 2\pi nt\,\frac{df}{dt} + \sin 2\pi nt\,\frac{d^2f}{dt^2}.$$

Putting $EI\frac{\pi^4}{l^4} = 4\pi^2 n_0^2 m$, where n_0 is the unloaded fundamental frequency of the girder and $ml = M_G$, the total mass of the girder, the equation for $f(t)$ takes the form

$$4\pi^2\left[n_0^2 - \frac{M}{M_G}n^2(1 - \cos 4\pi nt)\right]f(t)$$

$$+ 4\pi\left[n_b + n\frac{M}{M_G}\sin 4\pi nt\right]\frac{df}{dt} + \left[1 + \frac{M}{M_G}(1 - \cos 4\pi nt)\right]\frac{d^2f}{dt^2}$$

$$= \frac{P}{M_G}[\cos 2\pi(N-n)t - \cos 2\pi(N+n)t].$$

This equation can be solved by a series method somewhat similar to that employed in Chapter VII. In comparison with those methods, the introduction of damping adds somewhat to the labour of computing the forced oscillation, but on the other hand the series becomes more rapidly convergent and can consequently be limited to a shorter range. The determination of the free oscillation moreover need not be calculated with a high degree of precision, since damping causes these free oscillations to die out more or less rapidly, and they only figure prominently in the initial stages of the motion.

For the particular integral or forced oscillation, let

$$f(t) = A_1 \cos 2\pi (N+n) t + A_3 \cos 2\pi (N+3n) t + A_5 \cos 2\pi (N+5n) t + \ldots$$
$$+ A_{-1} \cos 2\pi (N-n) t + A_{-3} \cos 2\pi (N-3n) t + A_{-5} \cos 2\pi (N-5n) t + \ldots$$
$$+ B_1 \sin 2\pi (N+n) t + B_3 \sin 2\pi (N+3n) t + B_5 \cos 2\pi (N+5n) t + \ldots$$
$$+ B_{-1} \sin 2\pi (N-n) t + B_{-3} \sin 2\pi (N-3n) t + B_{-5} \sin 2\pi (N-5n) t + \ldots.$$

Write this in the abbreviated form

$$f(t) = \sum_{r=-\infty}^{r=+\infty} A_r \cos 2\pi q_r t + \sum_{r=-\infty}^{r=+\infty} B_r \sin 2\pi q_r t,$$

where $q_r = N + rn$, and r is any odd integer, positive or negative.

By substitution it will be found that the coefficient of $\cos 2\pi q_r t$ on the left-hand side of the equation for $f(t)$ is

$$4\pi^2 \left[A_r \left\{ n_0{}^2 - \frac{M}{M_G} n^2 - \left(1 + \frac{M}{M_G} \right) q_r{}^2 \right\} + \frac{M}{2M_G} A_{r-2} q^2{}_{r-1} \right.$$
$$\left. + \frac{M}{2M_G} A_{r+2} q^2{}_{r+1} + 2n_b q_r B_r \right],$$

and the coefficient for $\sin 2\pi q_r t$ is

$$4\pi^2 \left[B_r \left\{ n_0{}^2 - \frac{M}{M_G} n^2 - \left(1 + \frac{M}{M_G} \right) q_r{}^2 \right\} + \frac{M}{2M_G} B_{r-2} q^2{}_{r-1} \right.$$
$$\left. + \frac{M}{2M_G} B_{r+2} q^2{}_{r+1} - 2n_b q_r A_r \right].$$

Write these $2\pi^2 n_0{}^2 \dfrac{M}{M_G} [2A_r a_r + A_{r-2} b_{r-1} + A_{r+2} b_{r+1} + B_r c_r]$

and $2\pi^2 n_0{}^2 \dfrac{M}{M_G} [2B_r a_r + B_{r-2} b_{r-1} + B_{r+2} b_{r+1} - A_r c_r],$

where $a_r = \dfrac{M_G}{M} - \left(\dfrac{n}{n_0} \right)^2 - \left(1 + \dfrac{M}{M_G} \right) b_r,$

$$b_r = \left(\frac{q_r}{n_0} \right)^2,$$

$$c_r = \frac{4M_G}{M} \cdot \frac{n_b q_r}{n_0{}^2}.$$

Equating coefficients of $\cos 2\pi q_r t$ and $\sin 2\pi q_r t$ on both sides of the equation for $f(t)$, $[2A_r a_r + A_{r-2}b_{r-1} + A_{r+2}b_{r+1} + B_r c_r]$ has the value $\dfrac{P}{2\pi^2 n_0{}^2 M}$ when $r = -1$, it has the value $-\dfrac{P}{2\pi^2 n_0{}^2 M}$ when $r = +1$, and is zero for all other values of r. Similarly $[2B_r a_r + B_{r-2}b_{r-1} + B_{r+2}b_{r+1} - A_r c_r]$ is zero for all values of r.

From the former set of conditions the B coefficients can be expressed in terms of the A coefficients, and making use of the fact that

$$\frac{P}{2\pi^2 n_0{}^2 M_G} = D_P,$$

where D_P is the central deflection due to a steady central load P, the following series of equations for obtaining the A coefficients alone can be obtained:

$$A_r \left[\frac{4a_r{}^2}{c_r} + \frac{b^2{}_{r-1}}{c_{r-2}} + \frac{b^2{}_{r+1}}{c_{r+2}} + c_r \right] + A_{r+2} \left[2b_{r+1} \left(\frac{a_r}{c_r} + \frac{a_{r+2}}{c_{r+2}} \right) \right]$$
$$+ A_{r-2} \left[2b_{r-1} \left(\frac{a_r}{c_r} + \frac{a_{r-2}}{c_{r-2}} \right) \right] + A_{r+4} \left[\frac{b_{r+1}b_{r+3}}{c_{r+2}} \right] + A_{r-4} \left[\frac{b_{r-1}b_{r-3}}{c_{r-2}} \right]$$

has the value $\qquad -\dfrac{b_2}{c_1} \cdot \dfrac{M_G}{M} D_P$ when $r = 3$,

has the value $\qquad -\left[\dfrac{2a_1}{c_1} - \dfrac{b_0}{c_{-1}} \right] \dfrac{M_G}{M} D_P$ when $r = +1$,

has the value $\qquad +\left[\dfrac{2a_{-1}}{c_{-1}} - \dfrac{b_0}{c_1} \right] \dfrac{M_G}{M} D_P$ when $r = -1$,

has the value $\qquad +\dfrac{b_{-2}}{c_{-1}} \dfrac{M_G}{M} D_P$ when $r = -3$,

and is zero for all other values of r.

From this set of equations the A coefficients can be determined to any desired degree of accuracy, and when this has been done, the B coefficients can be deduced from the conditions that

$$[2A_r a_r + A_{r-2}b_{r-1} + A_{r+2}b_{r+1} + B_r c_r]$$

has the values $\dfrac{M_G}{M} D_P$ and $-\dfrac{M_G}{M} D_P$ when $r = -1$ and $+1$ respectively, and is zero for all other values of r.

This method will be illustrated by four numerical examples later on in this Chapter.

Free oscillation

The particular integral or forced oscillation obtained by the foregoing process does not satisfy either of the starting conditions, viz. $y = 0$ when $t = 0$ and $\dfrac{dy}{dt} = 0$ when $t = 0$. To represent the true state of oscillation built up from a condition of rest and zero deflection it is necessary to superpose a complementary or free oscillation of the type

$$y = f(t) \sin \frac{\pi x}{l},$$

where $f(t)$ is the solution of

$$4\pi^2 \left[n_0{}^2 - \frac{M}{M_G} n^2 (1 - \cos 4\pi nt) \right] f(t) + 4\pi \left[n_b + \frac{M}{M_G} n \sin 4\pi nt \right] \frac{df}{dt}$$

$$+ \left[1 + \frac{M}{M_G} (1 - \cos 4\pi nt) \right] \frac{d^2 f}{dt^2} = 0.$$

It is possible to effect a solution of this equation in the form of a series, using a process of trial and error somewhat similar to that adopted in Chapter VII for determining free oscillations. In this case, however, the existence of damping makes it unnecessary to resort to such an elaborate process. The free oscillations are relatively unimportant, in that they are only prominent at the early stages in the motion and they die down into insignificance by the time the load has reached the centre of the bridge. Accordingly, the following approximate representation of the free oscillation is sufficiently accurate for all practical purposes.

If during a short interval of time the coefficients of $f(t)$, $\dfrac{df}{dt}$ and $\dfrac{d^2 f}{dt^2}$, in the above differential equation for the free oscillation, are treated as constants, the solution takes the form

$$f(t) = e^{-\alpha t} [A \sin 2\pi N_0 t + B \cos 2\pi N_0 t],$$

where

$$\alpha = \frac{2\pi \left(n_b + n \dfrac{M}{M_G} \sin 4\pi nt \right)}{1 + \dfrac{2M}{M_G} \sin^2 2\pi nt},$$

$$N_0 = n_0 \sqrt{\frac{M_G}{M_G + 2M \sin^2 2\pi nt}};$$

and n^2 and $n_b{}^2$ are neglected in comparison with $n_0{}^2$.

This is equivalent to saying that when the load has moved a distance a along the bridge, the frequency of the free oscillations is given by

$$N_0 = n_0 \sqrt{\dfrac{M_G}{M_G + 2M \sin^2 \dfrac{\pi a}{l}}},$$

and, as such, is independent of the speed and damping, and for this position of the load the rate at which the amplitude of the oscillations is dying out is given by the multiplier

$$e^{-2\pi t \left[\left(n_b M_G + nM \sin \frac{2\pi a}{l} \right) / \left(M_G + 2M \sin^2 \frac{\pi a}{l} \right) \right]}.$$

From these two results the decrease in amplitude and the variation in the frequency of the free oscillations can be deduced with an accuracy which is sufficient for all practical purposes.

NUMERICAL EXAMPLES

The foregoing methods will now be illustrated by applying them to determine the oscillations and deflections set up in a long-span bridge by the passage of a locomotive.

The bridge and locomotive selected for this purpose are the same as those specified in the previous Chapter, but for convenience of reference their characteristics will be restated.

Bridge data.

Span $l = 270$ feet.

Total mass $M_G = 450$ tons.

Unloaded frequency $n_0 = 3$.

Damping coefficient $n_b = 0 \cdot 12$.

Locomotive data.

Total mass $M = 100$ tons.

Circumference of driving-wheels $= 15$ feet.

Hammer-blow at N revs. per sec. $= 0 \cdot 6 N^2$ tons.

Four cases will be considered:

Case (1) $N = 2$, $v = 30$ f.s.; a speed which is too low for resonance.

Case (2) $N = 2 \cdot 4961$, $v = 37 \cdot 4415$ f.s.; a speed which gives resonance when the locomotive is passing the centre of the bridge.

Case (3) $N = 2 \cdot 7136$, $v = 40 \cdot 7043$ f.s.; a speed which gives resonance when the locomotive is passing the quarter points of the bridge.

Case (4) $N = 6$, $v = 90$ f.s.; a high speed which is much too fast for resonance.

Case (1)

Forced oscillation

$$N=2, \quad v=30 \text{ f.s.,} \quad n=\frac{v}{2l}=\tfrac{1}{18},$$

$$P=0\cdot6N^2=2\cdot4 \text{ tons,}$$

$$q_r=N+rn=2+\tfrac{1}{18}r,$$

$$b_r=\left(\frac{q_r}{n_0}\right)^2=\frac{q_r{}^2}{9},$$

$$a_r=\frac{M_G}{M}-\left(\frac{n}{n_0}\right)^2-\left(1+\frac{M}{M_G}\right)b_r=4\cdot49966-5\cdot5b_r,$$

$$c_r=\frac{4M_G}{M}\cdot\frac{n_b q_r}{n_0{}^2}=0\cdot24q_r.$$

A preliminary investigation leads to the conclusion that a sufficiently accurate result is obtained if the series for $f(t)$ is limited within the range $r=-5$ to $r=+7$.

For laying down the equations by which the A and B coefficients are determined, the following table of the values of a_r, b_r and c_r is required:

r	a_r	b_r	c_r	r	a_r	b_r	c_r
0		0·4444		0		0·4444	
1	1·9175		0·4933	−1	2·1892		0·4667
2		0·4952		−2		0·3964	
3	1·6309		0·5200	−3	2·4456		0·4400
4		0·5487		−4		0·3512	
5	1·3291		0·5467	−5	2·6871		0·4133
6		0·6049					
7	1·0122		0·5733				

The equations for determining the seven A coefficients can now be laid down as follows:

$$8\cdot3910A_7 + 5\cdot0772A_5 + 0\cdot6071A_3 = 0,$$

$$5\cdot0772A_7 + 14\cdot6888A_5 + 6\cdot1098A_3 + 0\cdot5225A_1 = 0,$$

$$0\cdot6071A_7 + 6\cdot1098A_5 + 22\cdot0281A_3 + 6\cdot9560A_1 + 0\cdot4461A_{-1} + 4\cdot5173D_P = 0,$$

$$0\cdot5225A_5 + 6\cdot9560A_3 + 31\cdot2020A_1 + 7\cdot6240A_{-1} + 0\cdot3775A_{-3} + 30\cdot6988D_P = 0,$$

$$0\cdot4461A_3 + 7\cdot6240A_1 + 42\cdot3006A_{-1} + 8\cdot1254A_{-3} + 0\cdot3164A_{-5} - 38\cdot1634D_P = 0,$$

$$0\cdot3775A_1 + 8\cdot1254A_{-1} + 55\cdot4475A_{-3} + 8\cdot4707A_{-5} - 3\cdot8222D_P = 0,$$

$$0\cdot3164A_{-1} + 8\cdot4707A_{-3} + 70\cdot5751A_{-5} = 0.$$

From these, it will be found that:

$A_1 = -1\cdot3073 D_P$	$A_{-1} = 1\cdot1534 D_P$
$A_3 = 0\cdot1945 D_P$	$A_{-3} = -0\cdot0921 D_P$
$A_5 = -0\cdot0374 D_P$	$A_{-5} = 0\cdot0059 D_P$
$A_7 = 0\cdot0085 D_P$	

The corresponding B coefficients are then found from the equations

$$2\cdot0244 A_7 + 0\cdot6049 A_5 \qquad\qquad\quad + 0\cdot5733 B_7 = 0,$$

$$2\cdot6582 A_5 + 0\cdot5487 A_3 + 0\cdot6049 A_7 + 0\cdot5467 B_5 = 0,$$

$$3\cdot2618 A_3 + 0\cdot4952 A_1 + 0\cdot5487 A_5 + 0\cdot5200 B_3 = 0,$$

$$3\cdot8350 A_1 + 0\cdot4444 A_{-1} + 0\cdot4952 A_3 + 0\cdot4933 B_1 = -4\cdot5 D_P,$$

$$4\cdot3784 A_{-1} + 0\cdot3964 A_{-3} + 0\cdot4444 A_1 + 0\cdot4667 B_{-1} = +4\cdot5 D_P,$$

$$4\cdot8912 A_{-3} + 0\cdot3512 A_{-5} + 0\cdot3964 A_{-1} + 0\cdot4400 B_{-3} = 0,$$

$$5\cdot3742 A_{-5} \qquad\qquad\quad + 0\cdot3512 A_{-3} + 0\cdot4133 B_{-5} = 0.$$

From these it will be found that:

$B_1 = -0\cdot1932 D_P$	$B_{-1} = 0\cdot1444 D_P$
$B_3 = 0\cdot0642 D_P$	$B_{-3} = -0\cdot0201 D_P$
$B_5 = -0\cdot0230 D_P$	$B_{-5} = 0\cdot0018 D_P$
$B_7 = 0\cdot0093 D_P$	

Expressing the forced oscillation

$$y = [\Sigma A_r \cos 2\pi q_r t + \Sigma B_r \sin 2\pi q_r t] \sin \frac{\pi x}{l}$$

in the form

$$y = \cos 2\pi N t \left[\Sigma (A_r + A_{-r}) \cos 2\pi r n t + \Sigma (B_r - B_{-r}) \sin 2\pi r n t \right] \sin \frac{\pi x}{l}$$

$$+ \sin 2\pi N t \left[\Sigma (A_{-r} - A_r) \sin 2\pi r n t + \Sigma (B_r + B_{-r}) \cos 2\pi r n t \right] \sin \frac{\pi x}{l},$$

for the particular case under consideration

$$y = D_P \cos 2\pi N t \, [-0\cdot1539 \cos 2\pi n t + 0\cdot1024 \cos 6\pi n t - 0\cdot0315 \cos 10\pi n t$$

$$+ 0\cdot0085 \cos 14\pi n t - 0\cdot3382 \sin 2\pi n t + 0\cdot0843 \sin 6\pi n t$$

$$- 0\cdot0248 \sin 10\pi n t + 0\cdot0093 \sin 14\pi n t] \sin \frac{\pi x}{l}$$

$$+ D_P \sin 2\pi N t \, [2\cdot4607 \sin 2\pi n t - 0\cdot2866 \sin 6\pi n t + 0\cdot0432 \sin 10\pi n t$$

$$- 0\cdot0085 \sin 14\pi n t - 0\cdot0494 \cos 2\pi n t + 0\cdot0441 \cos 6\pi n t$$

$$- 0\cdot0212 \cos 10\pi n t + 0\cdot0093 \cos 14\pi n t] \sin \frac{\pi x}{l}.$$

Now D_P is the central deflection due to a steady central load of $2\cdot4$ tons; consequently if D_W denotes the central deflection due to the weight of the moving mass 100 tons, when standing at the centre of the bridge,

$$D_P = \frac{2\cdot4}{100} D_W.$$

Hence the connection between δ, the central deflection, and a, the distance the load has advanced along the bridge, is given by

$$\delta = \frac{D_W}{100} \cos 2\pi Nt \left[-0.3694 \cos \frac{\pi a}{l} + 0.2459 \cos \frac{3\pi a}{l} - 0.0756 \cos \frac{5\pi a}{l} \right.$$

$$+ 0.0205 \cos \frac{7\pi a}{l} - 0.8117 \sin \frac{\pi a}{l} + 0.2023 \sin \frac{3\pi a}{l}$$

$$\left. - 0.0596 \sin \frac{5\pi a}{l} + 0.0223 \sin \frac{7\pi a}{l} \right]$$

$$+ \frac{D_W}{100} \sin 2\pi Nt \left[5.9058 \sin \frac{\pi a}{l} - 0.6879 \sin \frac{3\pi a}{l} + 0.1038 \sin \frac{5\pi a}{l} \right.$$

$$- 0.0205 \sin \frac{7\pi a}{l} - 0.1186 \cos \frac{\pi a}{l} + 0.1058 \cos \frac{3\pi a}{l}$$

$$\left. - 0.0510 \cos \frac{5\pi a}{l} + 0.0223 \cos \frac{7\pi a}{l} \right].$$

Writing $\qquad \delta = \dfrac{D_W}{100} \left[\Phi\left(a\right) \cos 2\pi Nt + \Psi\left(a\right) \sin 2\pi Nt \right],$

where $N = 2$, the values of $\Phi\left(a\right)$ and $\Psi\left(a\right)$ in terms of a are as tabulated below:

a	0	$\frac{l}{16}$	$\frac{l}{8}$	$\frac{3l}{16}$	$\frac{l}{4}$	$\frac{5l}{16}$	$\frac{3l}{8}$	$\frac{7l}{16}$	$\frac{l}{2}$
$\Phi\left(a\right)$	-0.179	-0.270	-0.408	-0.575	-0.771	-1.009	-1.230	-1.294	-1.096
$\Psi\left(a\right)$	-0.042	0.784	1.642	2.562	3.524	4.502	5.462	6.280	6.618

a	l	$\frac{15l}{16}$	$\frac{7l}{8}$	$\frac{13l}{16}$	$\frac{3l}{4}$	$\frac{11l}{16}$	$\frac{5l}{8}$	$\frac{9l}{16}$	$\frac{l}{2}$
$\Phi\left(a\right)$	0.179	0.122	0.067	0.010	-0.037	-0.120	-0.338	-0.710	-1.096
$\Psi\left(a\right)$	0.042	0.888	1.783	2.725	3.738	4.824	5.860	6.573	6.718

Case (2)

Forced oscillation

$$N = 2.4961, \quad v = 37.4415 \, \text{f.s.}, \quad n = \frac{v}{2l} = 0.0693,$$

$$P = 0.6N^2 = 3.7383 \, \text{tons},$$

$$q_r = N + rn = 2.4961 + 0.0693r,$$

$$b_r = \left(\frac{q_r}{n_0}\right)^2 = \frac{q_r^2}{9},$$

$$a_r = \frac{M_G}{M} - \left(\frac{n}{n_0}\right)^2 - \left(1 + \frac{M}{M_G}\right) b_r = 4.4947 - 5.5 b_r,$$

$$c_r = \frac{4M_G}{M} \cdot \frac{n_b q_r}{n_0^2} = 0.24 q_r.$$

In this case a sufficiently close approximation to $f(t)$ is obtained, if the series $f(t) = \Sigma A_r \cos 2\pi q_r t + \Sigma B_r \sin 2\pi q_r t$ is limited within the range $r = -7$ to $r = +13$.

The values of a_r, b_r and c_r, required for the determination of the A and B coefficients, are tabulated as follows:

r	a_r	b_r	c_r	r	a_r	b_r	c_r
0		0·6923		0		0·6923	
1	0·4725		0·6157	−1	0·8955		0·5824
2		0·7713		−2		0·6175	
3	0·0265		0·6490	−3	1·2948		0·5492
4		0·8546		−4		0·5471	
5	−0·4432		0·6822	−5	1·6710		0·5159
6		0·9421		−6		0·4808	
7	−0·9366		0·7155	−7	2·0235		0·4826
8		1·0339					
9	−1·4536		0·7488				
10		1·1300					
11	−1·9937		0·7820				
12		1·2304					
13	−2·5576		0·8153				

The equations for determining the eleven A coefficients can now be laid down as follows:

$$-34\cdot8440A_{13} + 13\cdot9933A_{11} - 1\cdot7779A_9 = 0,$$
$$-13\cdot9933A_{13} + 24\cdot6758A_{11} - 10\cdot1490A_9 + 1\cdot5602A_7 = 0,$$
$$-1\cdot7779A_{13} + 10\cdot1490A_{11} - 15\cdot1628A_9 + 6\cdot7209A_7 - 1\cdot3613A_5 = 0,$$
$$-1\cdot5602A_{11} + 6\cdot7209A_9 - 8\cdot3481A_7 + 3\cdot6905A_5 - 1\cdot1802A_3 = 0,$$
$$-1\cdot3613A_9 + 3\cdot6905A_7 - 4\cdot1997A_5 + 1\cdot0406A_3 - 1\cdot0156A_1 = 0,$$
$$-1\cdot1802A_7 + 1\cdot0406A_5 - 2\cdot6901A_3 + 1\cdot2468A_1 - 0\cdot8673A_{-1}$$
$$= 5\cdot6372D_p,$$
$$-1\cdot0156A_5 - 1\cdot2468A_3 - 3\cdot8057A_1 - 3\cdot1915A_{-1} - 0\cdot7340A_{-3}$$
$$= 1\cdot5576D_p,$$
$$-0\cdot8673A_3 - 3\cdot1915A_1 - 7\cdot5627A_{-1} - 4\cdot8106A_{-3} - 0\cdot6151A_{-5}$$
$$= -8\cdot7786D_p,$$
$$-0\cdot7340A_1 - 4\cdot8107A_{-1} - 13\cdot9945A_{-3} - 6\cdot1238A_{-5} - 0\cdot5099A_{-7}$$
$$= -4\cdot7712D_p,$$
$$-0\cdot6151A_{-1} - 6\cdot1238A_{-3} - 23\cdot1894A_{-5} - 7\cdot1465A_{-7} = 0,$$
$$-0\cdot5099A_{-3} - 7\cdot1465A_{-5} - 34\cdot8681A_{-7} = 0.$$

From these, it will be found that:

$A_1 = -1\cdot7333D_p$	$A_{-1} = 2\cdot4136D_p$
$A_3 = -2\cdot2580D_p$	$A_{-3} = -0\cdot4185D_p$
$A_5 = 0\cdot4515D_p$	$A_{-5} = 0\cdot0476D_p$
$A_7 = 0\cdot8260D_p$	$A_{-7} = -0\cdot0036D_p$
$A_9 = 0\cdot4133D_p$	
$A_{11} = 0\cdot1370D_p$	
$A_{13} = 0\cdot0339D_p$	

The corresponding B coefficients are then found from the equations

$$-5{\cdot}1152A_{13} + 1{\cdot}2304A_{11} \qquad\qquad + 0{\cdot}8153B_{13} = 0,$$
$$-3{\cdot}9874A_{11} + 1{\cdot}1300A_9 \;\; + 1{\cdot}2304A_{13} + 0{\cdot}7820B_{11} = 0,$$
$$-2{\cdot}9072A_9 \;\; + 1{\cdot}0339A_7 \;\; + 1{\cdot}1300A_{11} + 0{\cdot}7480B_9 \;\; = 0,$$
$$-1{\cdot}8732A_7 \;\; + 0{\cdot}9421A_5 \;\; + 1{\cdot}0339A_9 \;\; + 0{\cdot}7155B_7 \;\; = 0,$$
$$-0{\cdot}8864A_5 \;\; + 0{\cdot}8546A_3 \;\; + 0{\cdot}9421A_7 \;\; + 0{\cdot}6822B_5 \;\; = 0,$$
$$+0{\cdot}0530A_3 \;\; + 0{\cdot}7715A_1 \;\; + 0{\cdot}8546A_5 \;\; + 0{\cdot}6490B_3 \;\; = 0,$$
$$+0{\cdot}9450A_1 \;\; + 0{\cdot}6923A_{-1} + 0{\cdot}7713A_3 \;\; + 0{\cdot}6157B_1 \;\; = -4{\cdot}5D_P,$$
$$+1{\cdot}7910A_{-1} + 0{\cdot}6175A_{-3} + 0{\cdot}6923A_1 \;\; + 0{\cdot}5824B_{-1} = +4{\cdot}5D_P,$$
$$+2{\cdot}5896A_{-3} + 0{\cdot}5471A_{-5} + 0{\cdot}6175A_{-1} + 0{\cdot}5492B_{-3} = 0,$$
$$+3{\cdot}3420A_{-5} + 0{\cdot}4808A_{-7} + 0{\cdot}5471A_{-3} + 0{\cdot}5159B_{-5} = 0,$$
$$+4{\cdot}0470A_{-7} + \qquad\qquad + 0{\cdot}4808A_{-5} + 0{\cdot}4826B_{-7} = 0.$$

From these, it will be found that

$B_1 = -4{\cdot}5335D_P$	$B_{-1} = 2{\cdot}8086D_P$
$B_3 = 1{\cdot}6498D_P$	$B_{-3} = -0{\cdot}7877D_P$
$B_5 = 2{\cdot}2747D_P$	$B_{-5} = 0{\cdot}1387D_P$
$B_7 = 0{\cdot}8145D_P$	$B_{-7} = -0{\cdot}0169D_P$
$B_9 = 0{\cdot}2577D_P$	
$B_{11} = 0{\cdot}0480D_P$	
$B_{13} = 0{\cdot}0061D_P$	

Expressing the forced oscillation in the form

$$y = \cos 2\pi N t \left[\Sigma\,(A_r + A_{-r})\cos 2\pi rnt + \Sigma\,(B_r - B_{-r})\sin 2\pi rnt\right]\sin\frac{\pi x}{l}$$

$$+ \sin 2\pi N t\left[\Sigma\,(A_{-r} - A_r)\sin 2\pi rnt + \Sigma\,(B_r + B_{-r})\cos 2\pi rnt\right]\sin\frac{\pi x}{l},$$

for the particular case under consideration

$$y = D_P\cos 2\pi N t\,[0{\cdot}6802\cos 2\pi nt - 2{\cdot}6766\cos 6\pi nt + 0{\cdot}4992\cos 10\pi nt$$
$$+ 0{\cdot}8223\cos 14\pi nt + 0{\cdot}4133\cos 18\pi nt + 0{\cdot}1370\cos 22\pi nt$$
$$+ 0{\cdot}0339\cos 26\pi nt - 7{\cdot}3422\sin 2\pi nt + 2{\cdot}4375\sin 6\pi nt$$
$$+ 2{\cdot}1360\sin 10\pi nt + 0{\cdot}8314\sin 14\pi nt + 0{\cdot}2577\sin 18\pi nt$$
$$+ 0{\cdot}0480\sin 22\pi nt + 0{\cdot}0061\sin 26\pi nt]\sin\frac{\pi x}{l}$$

$$+ D_P\sin 2\pi N t\,[4{\cdot}1469\sin 2\pi nt + 1{\cdot}8395\sin 6\pi nt - 0{\cdot}4039\sin 10\pi nt$$
$$- 0{\cdot}8296\sin 14\pi nt - 0{\cdot}4133\sin 18\pi nt - 0{\cdot}1370\sin 22\pi nt$$
$$- 0{\cdot}0339\sin 26\pi nt - 1{\cdot}7249\cos 2\pi nt + 0{\cdot}8622\cos 6\pi nt$$
$$+ 2{\cdot}4140\cos 10\pi nt + 0{\cdot}7976\cos 14\pi nt + 0{\cdot}2577\cos 18\pi nt$$
$$+ 0{\cdot}0480\cos 22\pi nt + 0{\cdot}0061\cos 26\pi nt]\sin\frac{\pi x}{l}.$$

If D_W denotes the central deflection due to the moving load of 100 tons, when it stands at the centre of the bridge,

$$D_P = \frac{3.7383}{100} D_W,$$

and the connection between δ, the central deflection, and a, the distance the load has advanced along the bridge, is given by

$$\delta = \frac{D_W}{100} \cos 2\pi Nt \left[2.5429 \cos \frac{\pi a}{l} - 10.0058 \cos \frac{3\pi a}{l} + 1.8660 \cos \frac{5\pi a}{l} \right.$$

$$+ 3.0741 \cos \frac{7\pi a}{l} + 1.5450 \cos \frac{9\pi a}{l} + 0.5121 \cos \frac{11\pi a}{l} + 0.1268 \cos \frac{13\pi a}{l}$$

$$- 27.4472 \sin \frac{\pi a}{l} + 9.1119 \sin \frac{3\pi a}{l} + 7.9850 \sin \frac{5\pi a}{l} + 3.1081 \sin \frac{7\pi a}{l}$$

$$\left. + 0.9633 \sin \frac{9\pi a}{l} + 0.1793 \sin \frac{11\pi a}{l} + 0.0229 \sin \frac{13\pi a}{l} \right]$$

$$+ \frac{D_W}{100} \sin 2\pi Nt \left[15.5023 \sin \frac{\pi a}{l} + 6.8765 \sin \frac{3\pi a}{l} - 1.5099 \sin \frac{5\pi a}{l} \right.$$

$$- 3.1013 \sin \frac{7\pi a}{l} - 1.5450 \sin \frac{9\pi a}{l} - 0.5121 \sin \frac{11\pi a}{l} - 0.1268 \sin \frac{13\pi a}{l}$$

$$- 6.4482 \cos \frac{\pi a}{l} + 3.2230 \cos \frac{3\pi a}{l} + 9.0220 \cos \frac{5\pi a}{l} + 2.9816 \cos \frac{7\pi a}{l}$$

$$\left. + 0.9633 \cos \frac{9\pi a}{l} + 0.1793 \cos \frac{11\pi a}{l} + 0.0229 \cos \frac{13\pi a}{l} \right].$$

Writing $\qquad \delta = \dfrac{D_W}{100} [\Phi(a) \cos 2\pi Nt + \Psi(a) \sin 2\pi Nt],$

where $N = 2.4961$, the values of $\Phi(a)$ and $\Psi(a)$ in terms of a are tabulated as follows:

a	0	$\frac{l}{16}$	$\frac{l}{8}$	$\frac{3l}{16}$	$\frac{l}{4}$	$\frac{5l}{16}$	$\frac{3l}{8}$	$\frac{7l}{16}$	$\frac{l}{2}$
$\Phi(a)$	-0.339	5.622	-0.673	-6.168	-9.649	-13.899	-19.312	-25.022	-30.875
$\Psi(a)$	9.944	2.179	-0.996	2.921	7.148	10.291	12.111	11.712	9.058

a	l	$\frac{15l}{16}$	$\frac{7l}{8}$	$\frac{13l}{16}$	$\frac{3l}{4}$	$\frac{11l}{16}$	$\frac{5l}{8}$	$\frac{9l}{16}$	$\frac{l}{2}$
$\Phi(a)$	0.339	15.382	12.535	-9.996	-30.386	-39.671	-40.371	-36.525	-30.875
$\Psi(a)$	-9.944	-1.107	22.766	34.481	28.292	16.926	9.062	7.038	9.058

Case (3)

Forced oscillation

$$N = 2 \cdot 7136, \quad v = 40 \cdot 7043 \text{ f.s.}, \quad n = \frac{v}{2l} = 0 \cdot 0754,$$

$$P = 0 \cdot 6N^2 = 4 \cdot 4182 \text{ tons},$$

$$q_r = N + rn = 2 \cdot 7136 + 0 \cdot 0754r,$$

$$b_r = \left(\frac{q_r}{n_0}\right)^2 = \frac{q_r^2}{9},$$

$$a_r = \frac{M_G}{M} - \left(\frac{n}{n_0}\right)^2 - \left(1 + \frac{M}{M_G}\right)b_r = 4 \cdot 4994 - 5 \cdot 5b_r,$$

$$c_r = \frac{4M_G}{M} \cdot \frac{n_b q_r}{n_0^2} = 0 \cdot 24q_r.$$

In this case a sufficiently close approximation to $f(t)$ is obtained, if the series $f(t) = \Sigma A_r \cos 2\pi q_r t + \Sigma B_r \sin 2\pi q_r t$ is limited within the range $r = -7$ to $r = +9$.

The values of a_r, b_r and c_r, required for the determination of the A and B coefficients, are tabulated as follows:

r	a_r	b_r	c_r	r	a_r	b_r	c_r
0		0·8182		0		0·8182	
1	−0·2543		0·6694	−1	0·2462		0·6332
2		0·9116		−2		0·7298	
3	−0·7823		0·7056	−3	0·7182		0·5970
4		1·0102		−4		0·6464	
5	−1·3378		0·7417	−5	1·1631		0·5608
6		1·1137		−6		0·5681	
7	−1·9213		0·7790	−7	1·5794		0·5246
8		1·2224					
9	−2·5329		0·8141				

The equations for determining the nine A coefficients can now be laid down as follows:

$$-34 \cdot 2573A_9 + 13 \cdot 6448A_7 - 1 \cdot 7501A_5 \qquad\qquad = 0,$$

$$-13 \cdot 6448A_9 + 23 \cdot 2672A_7 - 9 \cdot 5189A_5 + 1 \cdot 5169A_3 \qquad = 0,$$

$$- 1 \cdot 7501A_9 + 9 \cdot 5189A_7 - 13 \cdot 4344A_5 + 5 \cdot 8842A_3 - 1 \cdot 3051A_1 = 0,$$

$$- 1 \cdot 5169A_7 + 5 \cdot 8842A_5 - 6 \cdot 7923A_3 + 2 \cdot 7140A_1 - 1 \cdot 1142A_{-1}$$
$$= 6 \cdot 1282D_P,$$

$$- 1 \cdot 3051A_5 + 2 \cdot 7140A_3 - 3 \cdot 2908A_1 - 0 \cdot 0146A_{-1} - 0 \cdot 9430A_{-3}$$
$$= -9 \cdot 2338D_P,$$

$$- 1 \cdot 1142A_3 - 0 \cdot 0146A_1 - 2 \cdot 9083A_{-1} - 2 \cdot 3234A_{-3} - 0 \cdot 7902A_{-5}$$
$$= 2 \cdot 0008D_P,$$

$$- 0 \cdot 9430A_1 - 2 \cdot 3234A_{-1} - 5 \cdot 6392A_{-3} - 4 \cdot 2365A_{-5} - 0 \cdot 6548A_{-7}$$
$$= -5 \cdot 1865D_P,$$

$$- 0 \cdot 7902A_{-1} - 4 \cdot 2365A_{-3} - 11 \cdot 5250A_{-5} - 5 \cdot 7772A_{-7} \qquad = 0,$$

$$- 0 \cdot 6548A_{-3} - 5 \cdot 7772A_{-5} - 20 \cdot 1203A_{-7} \qquad\qquad = 0.$$

From these it will be found that:

$A_1 = 2\cdot7353D_P$	$A_{-1} = -2\cdot0875D_P$
$A_3 = 0\cdot4239D_P$	$A_{-3} = 1\cdot7137D_P$
$A_5 = -0\cdot1552D_P$	$A_{-5} = -0\cdot5360D_P$
$A_7 = -0\cdot1128D_P$	$A_{-7} = 0\cdot0981D_P$
$A_9 = -0\cdot0370D_P$	

The corresponding B coefficients are then found from the equations

$$-5\cdot0658A_9 \ +1\cdot2224A_7 \qquad\qquad\ +0\cdot8141B_9 \ =0,$$
$$-3\cdot8426A_7 \ +1\cdot1137A_5 \ +1\cdot2224A_9 \ +0\cdot7779B_7 \ =0,$$
$$-2\cdot6756A_5 \ +1\cdot0102A_3 \ +1\cdot1137A_7 \ +0\cdot7417B_5 \ =0,$$
$$-1\cdot5646A_3 \ +0\cdot9116A_1 \ +1\cdot0102A_5 \ +0\cdot7056B_3 \ =0,$$
$$-0\cdot5086A_1 \ +0\cdot8182A_{-1}+0\cdot9116A_3 \ +0\cdot6694B_1 \ =-4\cdot5D_P,$$
$$+0\cdot4924A_{-1}+0\cdot7298A_{-3}+0\cdot8182A_1 \ +0\cdot6332B_{-1}=+4\cdot5D_P,$$
$$+1\cdot4364A_{-3}+0\cdot6464A_{-5}+0\cdot7298A_{-1}+0\cdot5970B_{-3}=0,$$
$$+2\cdot3262A_{-5}+0\cdot5681A_{-7}+0\cdot6464A_{-3}+0\cdot5608B_{-5}=0,$$
$$+3\cdot1588A_{-7} \qquad\qquad\ +0\cdot5681A_{-5}+0\cdot5246B_{-7}=0.$$

From these, it will be found that:

$B_1 = -2\cdot6696D_P$	$B_{-1} = 3\cdot2205D_P$
$B_3 = -2\cdot3717D_P$	$B_{-3} = -0\cdot9909D_P$
$B_5 = -0\cdot9679D_P$	$B_{-5} = 0\cdot1487D_P$
$B_7 = -0\cdot2770D_P$	$B_{-7} = -0\cdot0104D_P$
$B_9 = -0\cdot0609D_P$	

Expressing the forced oscillation in the form

$$y = \cos 2\pi Nt\,[\Sigma\,(A_r+A_{-r})\cos 2\pi rnt + \Sigma\,(B_r-B_{-r})\sin 2\pi rnt]\sin\frac{\pi x}{l}$$
$$+\sin 2\pi Nt\,[\Sigma\,(A_{-r}-A_r)\sin 2\pi rnt + \Sigma\,(B_r+B_{-r})\cos 2\pi rnt]\sin\frac{\pi x}{l},$$

for this particular case

$$y = D_P\cos 2\pi Nt\,[0\cdot6478\cos 2\pi nt + 2\cdot1376\cos 6\pi nt - 0\cdot6912\cos 10\pi nt$$
$$-0\cdot0147\cos 14\pi nt - 0\cdot0370\cos 18\pi nt - 5\cdot8901\sin 2\pi nt - 1\cdot3808\sin 6\pi nt$$
$$-1\cdot1166\sin 10\pi nt - 0\cdot2666\sin 14\pi nt - 0\cdot0609\sin 18\pi nt]\sin\frac{\pi x}{l}$$
$$+D_P\sin 2\pi Nt\,[-4\cdot8228\sin 2\pi nt + 1\cdot2898\sin 6\pi nt - 0\cdot3808\sin 10\pi nt$$
$$+0\cdot2108\sin 14\pi nt + 0\cdot0370\sin 18\pi nt + 0\cdot5509\cos 2\pi nt - 3\cdot3626\cos 6\pi nt$$
$$-0\cdot8192\cos 10\pi nt - 0\cdot2874\cos 14\pi nt - 0\cdot0609\cos 18\pi nt]\sin\frac{\pi x}{l}.$$

If D_W denotes the central deflection due to the moving load of 100 tons, when it stands at the centre of the bridge,

$$D_P = \frac{4\cdot4182}{100}\,D_W,$$

and the connection between δ, the central deflection, and a, the distance the load has advanced along the bridge, is given by

$$\delta = \frac{D_W}{100} \cos 2\pi Nt \left[2 \cdot 8621 \cos \frac{\pi a}{l} + 9 \cdot 4443 \cos \frac{3\pi a}{l} - 3 \cdot 0539 \cos \frac{5\pi a}{l} \right.$$
$$- 0 \cdot 0649 \cos \frac{7\pi a}{l} - 0 \cdot 1635 \cos \frac{9\pi a}{l} - 22 \cdot 0236 \sin \frac{\pi a}{l} - 6 \cdot 1007 \sin \frac{3\pi a}{l}$$
$$\left. - 4 \cdot 9333 \sin \frac{5\pi a}{l} - 1 \cdot 1779 \sin \frac{7\pi a}{l} - 0 \cdot 2691 \sin \frac{9\pi a}{l} \right]$$
$$+ \frac{D_W}{100} \sin 2\pi Nt \left[- 21 \cdot 3081 \sin \frac{\pi a}{l} + 5 \cdot 6986 \sin \frac{3\pi a}{l} - 1 \cdot 6825 \sin \frac{5\pi a}{l} \right.$$
$$+ 0 \cdot 9314 \sin \frac{7\pi a}{l} + 0 \cdot 1635 \sin \frac{9\pi a}{l} + 2 \cdot 4340 \cos \frac{\pi a}{l} - 14 \cdot 8566 \cos \frac{3\pi a}{l}$$
$$\left. - 3 \cdot 6194 \cos \frac{5\pi a}{l} - 1 \cdot 2698 \cos \frac{7\pi a}{l} - 0 \cdot 2691 \sin \frac{9\pi a}{l} \right].$$

Writing
$$\delta = \frac{D_W}{100} \left[\Phi(a) \cos 2\pi Nt + \Psi(a) \sin 2\pi Nt \right],$$

where $N = 2 \cdot 7136$, the values of $\Phi(a)$ and $\Psi(a)$ in terms of a are tabulated as follows:

a	0	$\frac{l}{16}$	$\frac{l}{8}$	$\frac{3l}{16}$	$\frac{l}{4}$	$\frac{5l}{16}$	$\frac{3l}{8}$	$\frac{7l}{16}$	$\frac{l}{2}$
$\Phi(a)$	9·024	− 5·005	− 12·863	− 16·723	− 21·241	− 26·980	− 31·024	− 30·234	− 23·947
$\Psi(a)$	− 17·581	− 13·488	− 4·780	1·541	3·307	0·039	− 8·611	− 20·085	− 29·457

a	l	$\frac{15l}{16}$	$\frac{7l}{8}$	$\frac{13l}{16}$	$\frac{3l}{4}$	$\frac{11l}{16}$	$\frac{5l}{8}$	$\frac{9l}{16}$	$\frac{l}{2}$
$\Phi(a)$	− 9·024	− 22·970	− 28·139	− 23·679	− 15·928	− 10·607	− 10·296	− 15·585	− 23·947
$\Psi(a)$	17·581	10·856	− 3·520	− 16·516	− 24·089	− 28·732	− 32·416	− 33·486	− 29·457

Case (4)
Forced oscillation

$$N = 6, \quad v = 90 \text{ f.s.}, \quad n = \frac{v}{2l} = \tfrac{1}{6},$$
$$P = 0 \cdot 6N^2 = 21 \cdot 6 \text{ tons},$$
$$q_r = N + rn = 6 + \tfrac{1}{6}r,$$
$$b_r = \left(\frac{q_r}{n_0} \right)^2 = \frac{q_r^2}{9},$$
$$a_r = \frac{M_G}{M} - \left(\frac{n}{n_0} \right)^2 - \left(1 + \frac{M}{M_G} \right) b_r = 4 \cdot 4969 - 5 \cdot 5 b_r,$$
$$c_r = \frac{4M_G}{M} \cdot \frac{n_b q_r}{n_0^2} = 0 \cdot 24 q_r.$$

In this case a sufficiently close approximation to $f(t)$ is obtained, if the series $f(t) = \Sigma A_r \cos 2\pi q_r t + B_r \sin 2\pi q_r t$ is limited within the range $r = -7$ to $r = +7$.

The values of a_r, b_r and c_r, required for the determination of the A and B coefficients, are tabulated as follows:

r	a_r	b_r	c_r	r	a_r	b_r	c_r
0		4·0000		0		4·0000	
1	$-18·7423$		1·4800	-1	$-16·2978$		1·4000
2		4·4568		-2		3·5679	
3	$-21·3225$		1·5600	-3	$-13·9892$		1·3200
4		4·9383		-4		3·1605	
5	$-24·0386$		1·6400	-5	$-11·8164$		1·2400
6		5·4444		-6		2·7778	
7	$-26·8904$		1·7200	-7	$-9·7793$		1·1600

The equations for determining the eight A coefficients are as follows:

$$-1701·4071A_7 + 329·8395A_5 - 16·3940A_3 = 0,$$
$$-329·8395A_7 + 1443·9067A_5 - 279·7641A_3 + 14·1083A_1 = 0,$$
$$-16·3940A_7 + 279·7641A_5 - 1195·6177A_3 + 234·7127A_1$$
$$-12·0454A_{-1} = 13·5511D_P,$$
$$-14·1083A_5 + 234·7127A_3 - 975·0300A_1 + 194·4400A_{-1}$$
$$-10·1940A_{-3} = -126·8306D_P,$$
$$-12·0454A_3 + 194·4400A_1 - 780·7641A_{-1} + 158·6943A_{-3}$$
$$-8·5427A_{-5} = 116·9337D_P,$$
$$-10·1940A_1 + 158·6943A_{-1} - 611·4916A_{-3} + 127·2243A_{-5}$$
$$-7·0800A_{-7} = -11·4683D_P,$$
$$-8·5427A_{-1} + 127·2243A_{-3} - 465·8698A_{-5} + 99·7774A_{-7} = 0,$$
$$-7·0800A_{-3} + 99·7774A_{-5} - 337·1576A_{-7} = 0.$$

From these, it will be found that:

$A_1 = 0·10776D_P$	$A_{-1} = -0·12640D_P$
$A_3 = 0·01137D_P$	$A_{-3} = -0·01630D_P$
$A_5 = 0·00118D_P$	$A_{-5} = -0·00220D_P$
$A_7 = 0·00012D_P$	$A_{-7} = -0·00031D_P$

The corresponding B coefficients are then found from the equations

$$-53·7809A_7 + 5·4444A_5 + 1·7200B_7 = 0,$$
$$-48·0772A_5 + 4·9383A_3 + 5·4444A_7 + 1·6400B_5 = 0,$$
$$-42·6450A_3 + 4·4568A_1 + 4·9383A_5 + 1·5600B_3 = 0,$$
$$-37·4846A_1 + 4·0000A_{-1} + 4·4568A_3 + 1·4800B_1 = -4·5D_P,$$
$$-32·5956A_{-1} + 3·5679A_{-3} + 4·0000A_1 + 1·4000B_{-1} = +4·5D_P,$$
$$-27·9784A_{-3} + 3·1605A_{-5} + 3·5679A_{-1} + 1·3200B_{-3} = 0,$$
$$-23·6327A_{-5} + 2·7778A_{-7} + 3·1605A_{-3} + 1·2400B_{-5} = 0,$$
$$-19·5587A_{-7} + 2·7778A_{-5} + 1·1600B_{-7} = 0.$$

From these, it will be found that:

$B_1 = -0.00386D_P$	$B_{-1} = 0.00502D_P$
$B_3 = -0.00085D_P$	$B_{-3} = 0.00145D_P$
$B_5 = -0.00012D_P$	$B_{-5} = 0.00033D_P$
$B_7 = -0.00001D_P$	$B_{-7} = 0.00006D_P$

Expressing the forced oscillation in the form

$$y = \cos 2\pi Nt \left[\Sigma (A_r + A_{-r}) \cos 2\pi rnt + \Sigma (B_r - B_{-r}) \sin 2\pi rnt \right] \sin \frac{\pi x}{l}$$

$$+ \sin 2\pi Nt \left[\Sigma (A_{-r} - A_r) \sin 2\pi rnt + \Sigma (B_r + B_{-r}) \cos 2\pi rnt \right] \sin \frac{\pi x}{l},$$

for the particular case under consideration this takes the form

$$y = D_P \cos 2\pi Nt \left[-0.01864 \cos 2\pi nt - 0.00493 \cos 6\pi nt \right.$$

$$- 0.00102 \cos 10\pi nt - 0.00019 \cos 14\pi nt - 0.00888 \sin 2\pi nt$$

$$\left. - 0.00230 \sin 6\pi nt - 0.00045 \sin 10\pi nt - 0.00007 \sin 14\pi nt \right] \sin \frac{\pi x}{l}$$

$$+ D_P \sin 2\pi Nt \left[-0.23416 \sin 2\pi nt - 0.02767 \sin 6\pi nt \right.$$

$$- 0.00338 \sin 10\pi nt - 0.00043 \sin 14\pi nt + 0.00116 \cos 2\pi nt$$

$$\left. + 0.00060 \sin 6\pi nt + 0.00012 \sin 10\pi nt + 0.00005 \sin 14\pi nt \right] \sin \frac{\pi x}{l}.$$

If D_W denotes the central deflection due to the moving load, when it stands at the centre of the bridge,

$$D_P = \frac{21.6}{100} D_W,$$

and the connection between δ, the central deflection, and a, the distance the load has advanced along the bridge, is given by

$$\delta = \frac{D_W}{100} \cos 2\pi Nt \left[-0.4026 \cos \frac{\pi a}{l} - 0.1065 \cos \frac{3\pi a}{l} \right.$$

$$- 0.0220 \cos \frac{5\pi a}{l} - 0.0041 \cos \frac{7\pi a}{l} - 0.1918 \sin \frac{\pi a}{l}$$

$$\left. - 0.0497 \sin \frac{3\pi a}{l} - 0.0097 \sin \frac{5\pi a}{l} - 0.0015 \sin \frac{7\pi a}{l} \right]$$

$$+ \frac{D_W}{100} \sin 2\pi Nt \left[-5.0579 \sin \frac{\pi a}{l} - 0.5977 \sin \frac{3\pi a}{l} \right.$$

$$- 0.0730 \sin \frac{5\pi a}{l} - 0.0093 \sin \frac{7\pi a}{l} + 0.0251 \cos \frac{\pi a}{l}$$

$$\left. + 0.0130 \cos \frac{3\pi a}{l} + 0.0045 \cos \frac{5\pi a}{l} + 0.0011 \cos \frac{7\pi a}{l} \right].$$

Writing $\delta = \dfrac{D_W}{100}[\Phi(a)\cos 2\pi Nt + \Psi(a)\sin 2\pi Nt]$, where $N = 6$, the values of $\Phi(a)$ and $\Psi(a)$ in terms of a are tabulated as follows:

a	0	$\dfrac{l}{16}$	$\dfrac{l}{8}$	$\dfrac{3l}{16}$	$\dfrac{l}{4}$	$\dfrac{5l}{16}$	$\dfrac{3l}{8}$	$\dfrac{7l}{16}$	$\dfrac{l}{2}$
$\Phi(a)$	-0.535	-0.571	-0.530	-0.446	-0.359	-0.288	-0.230	-0.186	-0.150
$\Psi(a)$	0.044	-1.351	-2.533	-3.389	-3.935	-4.253	-4.424	-4.502	-4.524

a	l	$\dfrac{15l}{16}$	$\dfrac{7l}{8}$	$\dfrac{13l}{16}$	$\dfrac{3l}{4}$	$\dfrac{11l}{16}$	$\dfrac{5l}{8}$	$\dfrac{9l}{16}$	$\dfrac{l}{2}$
$\Phi(a)$	0.535	0.422	0.272	0.134	0.034	-0.034	-0.082	-0.118	-0.150
$\Psi(a)$	-0.044	-1.427	-2.584	-3.416	-3.947	-4.259	-4.426	-4.503	-4.524

Taking into account the deflection due to the weight of the locomotive, and adding the state of free oscillation necessary to satisfy the starting conditions, the connection between δ and a for these four cases is shewn by Fig. 26.

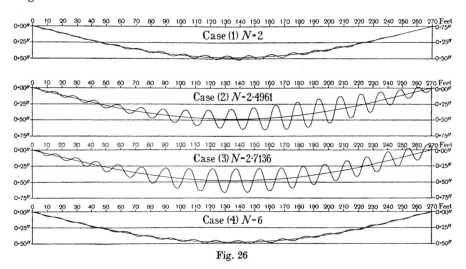

Fig. 26

The dynamic increments for Cases (2) and (3) are practically the same, thus indicating that the amplitude-frequency variation has no very sharply defined maximum, and, though not absolutely correct, there is no appreciable error in the statement that, to obtain the maximum dynamical deflection, the locomotive should be run at a speed which corresponds to the fully loaded frequency of the bridge, and that this maximum occurs when the locomotive is passing the centre of the span. If the hammer-blow frequency

EXPERIMENTAL DEFLECTION CURVES
Newark Dyke Bridge

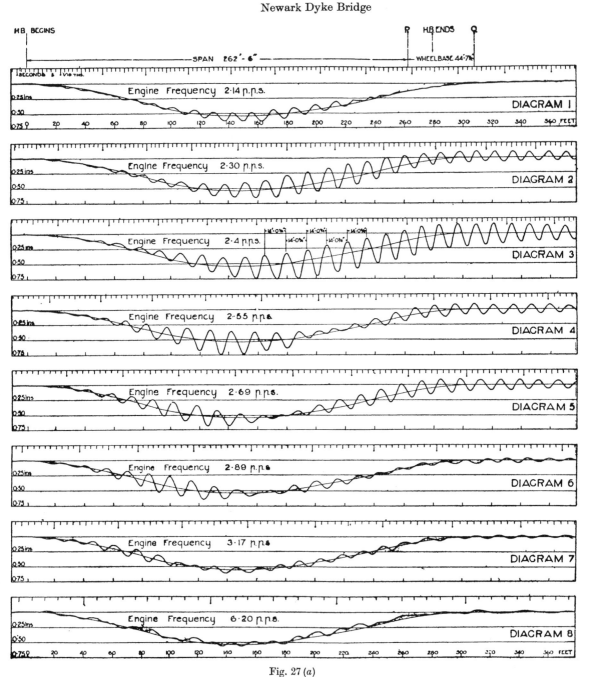

Fig. 27 (a)

THEORETICAL DEFLECTION CURVES
Newark Dyke Bridge

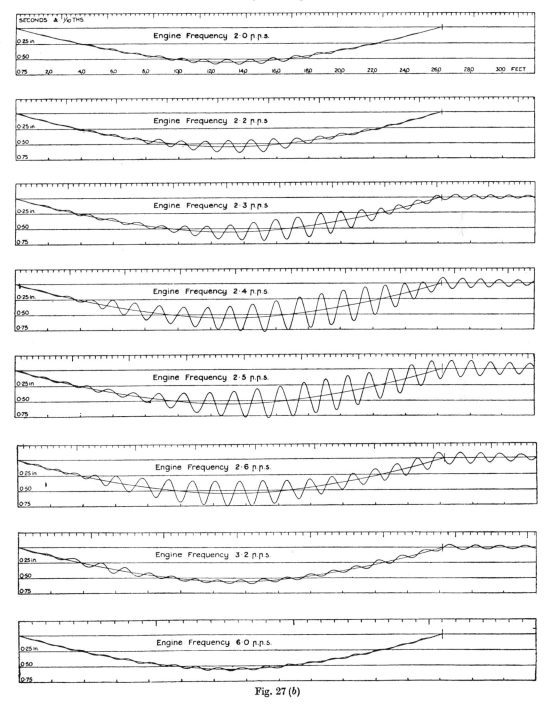

Fig. 27 (b)

is well below the range of loaded frequencies of the bridge, as in Case (1), the oscillations shew a marked reduction.

Case (4) illustrates what happens when the locomotive is going at full speed. In this case the hammer-blows reach the high value of 21·6 tons, but, owing to the fact that their frequency is then much above the range of loaded frequencies of the bridge, the oscillations developed by these large forces are relatively insignificant.

This set of calculations brings out quite clearly that, in the case of a long-span bridge, a speed limitation may be positively harmful, and, for this particular case, the dynamic increment of deflection at 60 m.p.h. is only about one-eighth of that at the comparatively slow speed of 25 m.p.h.

Confirmation of the accuracy of the results deduced by this full mathematical analysis is provided by the deflection diagrams shewn in Figs. 27 (a) and 27 (b), in which the experimental records and those given by theory can be directly compared.

The records relate to Newark Dyke Bridge, which is a single-track railway bridge, having a span of 262½ feet, a total mass of 460 tons, and an unloaded frequency of 2·88 p.p.s.

The locomotive used for the tests was an 0–8–0 coal locomotive, with wheels 14 feet circumference, and a hammer-blow of 0·576 ton at one rev. per sec. The total weight of the locomotive was 107·65 tons, and, when placed centrally, it lowered the natural frequency of the bridge to 2·4 p.p.s.

The locomotive was run at speeds ranging from 2·14 to 6·20 revs. per sec., and theoretical deflection diagrams more or less corresponding to this range of speeds were deduced. The general agreement is very satisfactory. Theory and experiment agree that the greatest oscillations occur when $N = 2·4$, and the maximum oscillations given by theory and experiment for this case are indistinguishable. The experimental records shew somewhat more pronounced residual oscillations than the theoretical results, but this can be accounted for by the fact that in the experiments there is evidence that the locomotive was slightly increasing speed during the run and so maintaining a more complete condition of resonance than theory contemplates.

In the insignificant character of the oscillations at speed well above the range of loaded frequencies, theory and experiment are again in agreement.

Although this general method yields results which are in excellent agreement with experiment, a little experience will convince the most lion-hearted computer that it is not a process of general utility, suitable for drawing-office practice. It is necessary, even at the cost of some slight loss of accuracy, to devise a simpler process for computing bridge oscillations, and a method of doing this will now be indicated. This simplified process is only applicable to the extreme case, namely the case when synchronism is established at the moment the locomotive is passing the centre of the bridge, but this is the only information the bridge designer really requires, since it enables him to deduce the maximum dynamic loads the bridge has to support and the consequential dynamic allowances which have to be provided.

APPROXIMATE METHOD FOR DETERMINING THE MAXIMUM STATE OF OSCILLATION IN A LONG-SPAN BRIDGE

The complication of the full mathematical analysis arises from the fact that the coefficients in the differential equation involved are not constants, and this in turn is due to the fact that the loaded frequency of the bridge varies with the position of the load. If the frequency of the bridge can be assumed constant, these complications disappear, and the solution of the differential equation presents no difficulty.

To obtain an approximation to the case when synchronism occurs at the instant the locomotive is passing the centre of the span, it will be assumed that the bridge has this fully loaded frequency for all positions of the load. In other words it will be assumed that its fundamental natural frequency is decreased to $n_0' = n_0 \sqrt{\dfrac{M_G}{M_G + 2M}}$, the inertia of the locomotive being taken into account by the addition of a mass $2M$, uniformly distributed along the bridge.

Under these circumstances, if $N = n_0' = n_0 \sqrt{\dfrac{M_G}{M_G + 2M}}$, resonance will be established for the whole period of the run, instead of momentarily at the centre.

This assumption of constant bridge frequency obviously tends to exaggerate the state of oscillation, but the presence of the inevitable bridge damping makes this over-statement remarkably small.

The change in the effective mass of the bridge necessitates a corresponding modification in the value of the damping coefficient n_b.

The damping resistance per unit length of bridge was taken to be $4\pi n_b m \dfrac{dy}{dt}$. If m is increased in the ratio $\dfrac{M_G + 2M}{M_G}$, then in order to keep the damping resistance unchanged, n_b must be decreased to a corresponding extent and the modified damping coefficient to be employed is

$$n_b' = n_b \frac{M_G}{M_G + 2M}.$$

With these modified bridge characteristics, viz.

$$M_G' = M_G + 2M, \quad n_0' = n_0 \sqrt{\frac{M_G}{M_G + 2M}}, \quad \text{and} \quad n_b' = n_b \frac{M_G}{M_G + 2M},$$

the hammer-blow effect on the bridge is then treated as being due to an alternating force $P \sin 2\pi N t$, moving with velocity v along the span, and the mathematical treatment of this case is precisely the same as was developed in Chapter VI.

For the forced oscillation, a very close approximation is obtained in the form

$$y = \frac{P}{4\pi^2 N^2 M_G'} \cdot \frac{N}{n^2 + n_b'^2} [n \cos 2\pi n t - n_b' \sin 2\pi n t] \cos 2\pi N t \sin \frac{\pi x}{l},$$

where $n = \dfrac{v}{2l}$.

To satisfy the starting conditions, a state of free oscillation of the form

$$y = A e^{-2\pi n_b' t} \cos 2\pi N t \sin \frac{\pi x}{l}$$

must be superposed. Hence for the synchronous case, where

$$N = n_0' = n_0 \sqrt{\frac{M_G}{M_G + 2M}},$$

the complete state of oscillation built up by an alternating force $P \sin 2\pi N t$ as it moves with velocity v along the bridge is given by

$$y = \frac{P}{4\pi^2 N^2 M_G'} \cdot \frac{N}{n^2 + n_b'^2} [n (\cos 2\pi n t - e^{-2\pi n_b' t}) - n_b' \sin 2\pi n t] \cos 2\pi N t \sin \frac{\pi x}{l}.$$

Suppose that the alternating force is associated with a steady gravity force W, moving with it along the bridge, and suppose that the central deflection produced by W, when placed stationary at the centre of the bridge, is D_W; then, since

$$\frac{P}{4\pi^2 N^2 M_G'} = \frac{P}{4\pi^2 n_0'^2 M_G'} = \frac{P}{2W} D_W,$$

the state of oscillation produced by the alternating force can be written in the form

$$y = D_W \frac{P}{2W} \cdot \frac{N}{n^2 + n_b'^2} [n (\cos 2\pi n t - e^{-2\pi n_b' t}) - n_b' \sin 2\pi n t] \cos 2\pi N t \sin \frac{\pi x}{l},$$

where P and W must be expressed in the same units, say tons.

By the time the load has reached the centre of the bridge, the free oscillations associated with the term $e^{-2\pi n_b' t}$ have been damped into insignificance.

At this instant $\cos 2\pi nt = 0$ and $\sin 2\pi nt = 1$. Hence the state of oscillation which exists when the load is passing the centre of the span is defined by

$$y = -D_W \frac{P}{2W} \cdot \frac{N n_b'}{n^2 + n_b'^2} \cos 2\pi Nt \sin \frac{\pi x}{l}.$$

The greatest oscillation occurs at a somewhat later stage, namely when $n \cos 2\pi nt - n_b' \sin 2\pi nt$ is a maximum, that is, when the load has passed the centre of the bridge by a distance $\dfrac{l}{\pi} \tan^{-1} \dfrac{n}{n_b'}$, and the semi-amplitude of this extreme oscillation is

$$D_W \frac{P}{2W} \cdot \frac{N}{\sqrt{n^2 + n_b'^2}}.$$

These approximate results will now be tested for accuracy against Case (2) of the oscillations represented in Fig. 26.

In this case
$$n_0' = n_0 \sqrt{\frac{M_G}{M_G + 2M}} = 3\sqrt{\frac{450}{650}} = 2 \cdot 4961,$$

$$n_b' = n_b \frac{M_G}{M_G + 2M} = 0 \cdot 12 \times \frac{450}{650} = 0 \cdot 0831,$$

$$N = n_0' = 2 \cdot 4961, \quad n = \frac{v}{2l} = 0 \cdot 0693,$$

$$P = 0 \cdot 6N^2 = 0 \cdot 6 \times (2 \cdot 4961)^2 = 3 \cdot 7383 \text{ tons,}$$

$$\frac{P}{2W} \cdot \frac{Nn}{n^2 + n_b'^2} = \frac{27 \cdot 66}{100},$$

$$\frac{P}{2W} \cdot \frac{N n_b'}{n^2 + n_b'^2} = \frac{33 \cdot 11}{100}.$$

Hence the central deflection δ for the oscillation produced by the hammer-blow, expressed in terms of a, the distance along the bridge traversed by the load, is given by

$$\delta = \frac{D_W}{100}\left[27 \cdot 66 \left(\cos \frac{\pi a}{l} - e^{-0 \cdot 5215 \frac{a}{v}} \right) - 33 \cdot 11 \sin \frac{\pi a}{l} \right] \cos 2\pi Nt.$$

The central deflection due to gravity force of 100 tons moving along the bridge is given by

$$\delta = \frac{D_W}{100}\left[100 \sin \frac{\pi a}{l} - 2 \cdot 78 e^{-0 \cdot 5215 \frac{a}{v}} \sin 2\pi Nt \right].$$

The result of plotting δ in terms of a, for the combined action of the steady and alternating force, is shewn in Fig. 28, and, for comparison, the corresponding Case (2), worked out by the full mathematical analysis, is shewn above.

The general agreement between these two diagrams is quite remarkably close.

When the locomotive is at the centre of the bridge, the approximate expression for the dynamical increment of deflection, viz. $D_W \dfrac{P}{2W} \cdot \dfrac{N n_b{}'}{n^2 + n_b{}'^2}$, gives $\dfrac{33 \cdot 11}{100} D_W$ as against $\dfrac{32 \cdot 18}{100} D_W$ by the exact method.

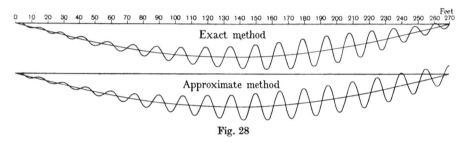

Fig. 28

The greatest increase of the central deflection, deduced by the approximate method, is $\dfrac{35 \cdot 95}{100} D_W$, as against $\dfrac{35 \cdot 30}{100} D_W$, given by the exact analysis. The maximum oscillation obtained by the approximate process has a semi-amplitude $\dfrac{43 \cdot 16}{100} D_W$, as compared with the exact estimate of $\dfrac{43 \cdot 13}{100} D_W$.

These results amply justify the use of the approximate method for determining the maximum state of oscillation in a long-span bridge.

SUMMARY

For calculating the dynamical increment of central deflection in a long-span bridge due to the hammer-blow of a locomotive, it is most convenient to express this in terms of the central deflection produced by the hammer-blow when treated as a statical force acting at the centre of the bridge, and, proceeding thus, the following important yet comparatively simple formulae are obtained.

For a long-span single-track bridge of length l, total mass M_G, unloaded natural frequency n_0, damping coefficient n_b, traversed by a locomotive of mass M and hammer-blow $P \sin 2\pi N t$, the maximum state of oscillation is developed when N, the revs. per sec. of the driving-wheels, has the value

$$N = n_0 \sqrt{\dfrac{M_G}{M_G + 2M}},$$ and the corresponding maximum central dynamic

deflection is given by the formula

$$\delta_0 = \frac{D_P}{2} \frac{N}{\sqrt{n^2 + n_b'^2}},$$

where $n = \dfrac{v}{2l}$, $n_b' = n_b \dfrac{M_G}{M_G + 2M}$, and D_P is the central deflection due to a force P statically applied at the centre of the bridge.

This maximum dynamic deflection occurs when the locomotive has passed the centre of the span by a distance

$$\frac{l}{\pi} \tan^{-1} \frac{n}{n_b'}.$$

The value of M which enters into the formula for determining N is to be taken as the mass of the locomotive excluding its tender. As will be explained later, the tender, being comparatively free on its springs, has little effect on the natural frequency of the bridge. It may lower the frequency to a small extent, but its dynamical effect on the bridge is mainly apparent in a slight increase of damping, and, neglecting the influence of the tender, so far as it introduces any inaccuracy, gives an error on the side of safety.

For double-heading on a single-track long-span bridge with two similar locomotives each of mass M and hammer-blow $P \sin 2\pi N t$, their longitudinal spacing being d and their hammer-blows being exactly in step, the maximum central dynamic deflection is given by the formula

$$\delta_0 = D_P \frac{N}{\sqrt{n^2 + n_b'^2}} \cos \frac{\pi d}{2l},$$

where, in this case,

$$N = n_0 \sqrt{\frac{M_G}{M_G + 4M \cos^2 \frac{\pi d}{2l}}},$$

$$n_b' = n_b \frac{M_G}{M_G + 4M \cos^2 \frac{\pi d}{2l}},$$

and D_P, again, is the central deflection due to a force P statically applied at the centre of the bridge.

For a double-track long-span bridge with two similar locomotives, one on each track passing one another at the centre of the span, the maximum central dynamic deflection is given by

$$\delta_0 = D_P \frac{N}{\sqrt{n^2 + n_b'^2}},$$

where, in this case,
$$N = n_0 \sqrt{\frac{M_G}{M_G + 4M}},$$

and
$$n_b{}' = n_b \frac{M_G}{M_G + 4M},$$

D_P being the same as in the preceding cases.

For double-heading on a double-track long-span bridge, with four similar locomotives two on each track, the greatest conceivable central dynamic deflection is given by

$$\delta_0 = 2D_P \frac{N}{\sqrt{n^2 + n_b{}'^2}} \cos \frac{\pi d}{2l},$$

where
$$N = n_0 \sqrt{\frac{M_G}{M_G + 8M \cos^2 \frac{\pi d}{2l}}}, \quad n_b{}' = n_b \frac{M_G}{M_G + 8M \cos^2 \frac{\pi d}{2l}},$$

and D_P, again, is the central deflection due to a force P statically applied at the centre of the bridge.

This last case assumes that both pairs of locomotives have the same speed and that the four hammer-blows are all exactly in step, a combination of circumstances which is so unlikely that it is hardly necessary to make provision in full for this almost inconceivable state of affairs.

When the locomotives are at the centre of the bridge in the case of single-heading, and equally spaced on either side of the centre in the case of double-heading, the central dynamic deflection is only slightly less than the maximum values stated above, and, even when the locomotives have reached the end of the bridge, the central dynamic deflection may still be a large percentage of the maximum. Consequently, for the sake of simplicity and to ensure a margin of safety, in computing dynamic allowances it is advisable to assume that the maximum state of oscillation is achieved by the time the locomotives have reached their central position and remains constant for the remainder of their passage across the bridge.

This prolonged state of maximum oscillation produced by isolated locomotives, if superposed upon the "crawl-deflection" for these same locomotives when followed by their trains, will over-state dynamic effects to some small extent, since, as in the case of a tender, the presence of a train tends to damp down the oscillations due to a locomotive. The over-statement, however, is too small to be of practical importance, and the error, such as it is, being on the side of safety, may even be regarded as beneficial.

CHAPTER IX

OSCILLATIONS TAKING INTO ACCOUNT THE SPRING MOVEMENT OF A LOCOMOTIVE

In the foregoing Chapters it has been assumed that a locomotive, in moving across a bridge, behaves as though the springs were locked. For long-span bridges their low natural frequency makes this assumption justifiable, since, for such bridges, the oscillations induced, when large in magnitude, are low in frequency, and the vertical accelerations they impose on a locomotive are not sufficient to break down the large amount of friction inherent in its spring mechanism. Thus for a bridge of say 250 feet span, oscillations with a semi-amplitude of 1/5 of an inch and a frequency of 2·5 are typical of the maximum which may be expected, and the vertical acceleration associated with this movement is about 4 f.s. units. If the spring-borne mass of the locomotive is 80 tons, the friction force required to hold the springs locked is about 10 tons, and in practice it is found that the friction force resisting spring movement is of this order of magnitude. In comparison, for a medium length bridge of say 150 feet span, although the maximum oscillations may possibly be halved, their frequency may well be doubled, and, as a consequence, the force required to hold the springs locked mounts up to 20 tons, which is considerably in excess of what spring friction can provide.

To be precise, if the spring-borne mass of the locomotive is M_S tons, and F tons is the total frictional resistance in the springing mechanism, spring movement will be induced, if at any instant δ, the semi-amplitude of the oscillations measured in feet, is such that

$$\delta > \frac{Fg}{4\pi^2 N^2 M_S}.$$

For bridges up to about 200 feet span, spring movement is to be expected, and as will be shewn later on, this phenomenon has a profound effect on bridge oscillations and raises the apparent natural frequency of the bridge to such an extent that the loaded frequency may actually be higher than the frequency when unloaded. The fact that synchronism may be looked for at high engine speeds associated with large hammer-blows would suggest that in medium-span bridges exceptionally violent oscillations may be anticipated, but this danger is fortunately counteracted to a considerable extent by the extra damping influence introduced by spring friction, which exercises a potent influence in checking the exuberance of these high speed synchronous oscillations.

The full mathematical analysis for calculating bridge oscillations due to a moving locomotive, taking into account spring movement, though practicable, is very laborious. It is useful as an occasional check, but for practical purposes a simpler method of prediction must be adopted, and one which gives away but little in accuracy can be devised as follows. Owing to the heavy damping introduced by spring movement, free oscillations subside rapidly, and this means that the oscillations mount up very quickly to the limit which damping prescribes. Thus, although the time during which the locomotive is in the central region of the bridge may be brief, it is nevertheless sufficient to allow the oscillations to approximate very nearly to the magnitude they would attain if the locomotive continued to hammer on the bridge at its centre for an indefinite length of time. In other words, the maximum state of oscillation can be estimated approximately by treating the locomotive as an oscillator fixed at the centre of the bridge, the locomotive skidding its wheels at a constant speed but having no motion of translation. This method of predicting the maximum bridge oscillations corresponding to a given engine speed may, as will be shewn later, be in error to an extent of possibly 15 per cent., but the error has the advantage that it is an over-statement and as such is in the nature of a beneficial safeguard. This simplified process is only applicable to bridges of medium span in which damping is heavily augmented by spring movement. In long-span bridges damping is small, free oscillations subside slowly, and in consequence the oscillations produced by a moving locomotive have not sufficient time to rise to anything approaching the value they would ultimately attain under the influence of a stationary skidding locomotive. For long-span bridges the skidding locomotive method of predicting the maximum state of oscillation leads to values which may be 50 per cent. or more in excess of the correct estimate, and as such it serves no useful purpose.

OSCILLATIONS PRODUCED BY A LOCOMOTIVE SKIDDING ITS WHEELS AT THE CENTRE OF A BRIDGE

For the purpose of analysis the locomotive is idealized as shewn in Fig. 29. The notation adopted is as follows:

Bridge data.

M_G denotes the total mass.

m ,,   the mass per unit length.

l ,,   the span.

n_0 ,,   the unloaded fundamental frequency.

n_b ,,   the damping coefficient.

Locomotive data.

M_S denotes the spring-borne mass (excluding the tender).

M_U denotes the unsprung mass.

$P \sin 2\pi Nt$ is the hammer-blow.

n_s denotes the natural frequency of the locomotive on its springs.

n_d is a coefficient specifying the damping due to the springs.

At any instant t during an oscillation, let h and z be the respective depths of M_U and M_S below their mean positions. Then no matter what is the nature or extent of the friction in the spring movement, the total reaction between the wheels and the rails is given by

$$P \sin 2\pi Nt - M_U \frac{d^2h}{dt^2} - M_S \frac{d^2z}{dt^2}.$$

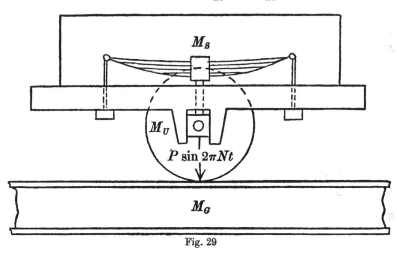

Fig. 29

The primary harmonic component of this central force is

$$\frac{2}{l}\left[P \sin 2\pi Nt - M_U \frac{d^2h}{dt^2} - M_S \frac{d^2z}{dt^2} \right] \sin \frac{\pi x}{l},$$

and the primary component of the bridge oscillation is given by the equation

$$EI \frac{d^4y}{dx^4} + 4\pi n_b m \frac{dy}{dt} + m \frac{d^2y}{dt^2} = \frac{2}{l}\left[P \sin 2\pi Nt - M_U \frac{d^2h}{dt^2} - M_S \frac{d^2z}{dt^2} \right] \sin \frac{\pi x}{l},$$

where $4\pi n_b m \dfrac{dy}{dt}$ is the resistance per unit length of span which accounts for damping.

For the state of steady oscillation, let $y = f(t) \sin \dfrac{\pi x}{l}$. Then $h = f(t)$, and the differential equation for $f(t)$ takes the form

$$EI \frac{\pi^4}{l^4} f(t) + 4\pi n_b m \frac{df}{dt} + m \frac{d^2f}{dt^2} = \frac{2}{l}\left[P \sin 2\pi Nt - M_U \frac{d^2f}{dt^2} - M_S \frac{d^2z}{dt^2} \right],$$

and, since $EI\dfrac{\pi^4}{l^4}=4\pi^2n_0{}^2m$, the equation can be written

$$4\pi^2n_0{}^2M_G f(t)+4\pi n_b\, M_G\frac{df}{dt}+(M_G+2M_U)\frac{d^2f}{dt^2}=2P\sin 2\pi Nt-2M_S\frac{d^2z}{dt^2}$$

$$\dots\dots(1).$$

If the springing is so perfect that $\dfrac{d^2z}{dt^2}=0$, the inertia effect of the locomotive merely amounts to an additional mass M_U at the centre of the bridge. If, on the other hand, friction is so great that the springs are locked, $\dfrac{d^2z}{dt^2}=\dfrac{d^2f}{dt^2}$, and the inertia effect of the locomotive amounts to an additional mass M_U+M_S at the centre of the span.

The latter state of affairs, as has been previously mentioned, is associated with long-span bridges in which the comparatively sluggish oscillations are not violent enough to break down spring friction. For bridges of moderate span the term $2M_S\dfrac{d^2z}{dt^2}$, which appears on the right-hand side of the equation for $f(t)$, exercises a profound effect on the amplitude of the bridge oscillations and the frequencies at which large oscillations are developed.

By considering the motion of M_S another differential equation can be established connecting z with $f(t)$, and this, unlike the previous relationship, calls for a knowledge of the strength of the springs and the nature of the damping influences associated with spring movement.

Since z and $f(t)$ are the depths of M_S and M_U below their mean positions, the extra compression in the springs is $z-f(t)$, and the extra upward thrust in the springs can be written $\mu[z-f(t)]$.

In the first instance, assume that the resistance to spring movement is of the fluid friction variety depending on the relative velocity of M_S and M_U. This force acting upwards on M_S will accordingly be written $K\left[\dfrac{dz}{dt}-\dfrac{df}{dt}\right]$ and the equation of motion for M_S is

$$-M_S\frac{d^2z}{dt^2}=\mu[z-f(t)]+K\left[\frac{dz}{dt}-\frac{df}{dt}\right]$$

or $\qquad\qquad M_S\dfrac{d^2z}{dt^2}+K\dfrac{dz}{dt}+\mu z=\mu f(t)+K\dfrac{df}{dt}.$

For a free, frictionless, oscillation of M_S on its springs, M_U remaining stationary,

$$M_S\frac{d^2z}{dt^2}+\mu z=0,$$

and, if n_s is the frequency of this oscillation,

$$4\pi^2 n_s^2 = \frac{\mu}{M_s}.$$

For the corresponding oscillation when friction is taken into account

$$M_s \frac{d^2z}{dt^2} + K \frac{dz}{dt} + \mu z = 0.$$

A solution of this can be obtained in the form

$$z = A e^{-\frac{K}{2M_s}t} \sin \left[\frac{\mu}{M_s} - \left(\frac{K}{2M_s}\right)^2\right]^{\frac{1}{2}} t.$$

Writing $\dfrac{K}{2M_s} = 2\pi n_d$, where n_d has the dimensions of a frequency, the expression for z takes the simpler form

$$z = A e^{-2\pi n_d t} \sin 2\pi (n_s^2 - n_d^2)^{\frac{1}{2}} t,$$

and the equation of motion for M_s due to oscillations of the girder can be written in the form

$$\frac{d^2z}{dt^2} + 4\pi n_d \frac{dz}{dt} + 4\pi^2 n_s^2 z = 4\pi^2 n_s^2 f(t) + 4\pi n_d \frac{df}{dt} \qquad ......(2).$$

Combining this with equation (1) previously established, a differential equation for determining $f(t)$ can now be obtained as follows.

From equation (2)

$$\frac{d^4z}{dt^4} + 4\pi n_d \frac{d^3z}{dt^3} + 4\pi^2 n_s^2 \frac{d^2z}{dt^2} = 4\pi^2 n_s^2 \frac{d^2f}{dt^2} + 4\pi n_d \frac{d^3f}{dt^3}.$$

From equation (1)

$$\frac{d^2z}{dt^2} = \frac{1}{2M_s}\left[2P \sin 2\pi Nt - 4\pi^2 n_0^2 M_G f(t) - 4\pi n_b m \frac{df}{dt} - (M_G + 2M_U)\frac{d^2f}{dt^2}\right].$$

Substituting in the first of these two equations the values of $\dfrac{d^2z}{dt^2}, \dfrac{d^3z}{dt^3}$ and $\dfrac{d^4z}{dt^4}$ obtained from the second, the equation for $f(t)$ takes the form

$$(M_G + 2M_U)\frac{d^4f}{dt^4} + 4\pi [n_b M_G + n_d (M_G + 2M_U + 2M_s)]\frac{d^3f}{dt^3}$$

$$+ 4\pi^2 [n_0^2 M_G + 4 n_b n_d M_G + n_s^2 (M_G + 2M_U + 2M_s)]\frac{d^2f}{dt^2}$$

$$+ 16\pi^3 (n_0^2 n_d M_G + n_s^2 n_b M_G)\frac{df}{dt} + 16\pi^4 n_0^2 n_s^2 M_G f(t)$$

$$= 8\pi^2 (n_s^2 - N^2) P \sin 2\pi Nt + 16\pi^2 N n_d P \cos 2\pi Nt.$$

If n_1 is the fundamental frequency of the girder when it carries an additional mass M_U at its centre, and n_2 is the fundamental frequency of the girder when it carries an additional mass $M_U + M_S$ at its centre,

$$\frac{n_1{}^2}{n_0{}^2} = \frac{M_G}{M_S + 2M_U} \quad \text{and} \quad \frac{n_2{}^2}{n_0{}^2} = \frac{M_G}{M_G + 2M_U + 2M_S}.$$

Hence the equation for $f(t)$ can be reduced to the form

$$\frac{d^4f}{dt^4} + 2\pi \left(2n_b \frac{n_1{}^2}{n_0{}^2} + 2n_d \frac{n_1{}^2}{n_2{}^2} \right) \frac{d^3f}{dt^3} + 4\pi^2 \left(n_1{}^2 + 4n_b n_d \frac{n_1{}^2}{n_0{}^2} + n_s{}^2 \frac{n_1{}^2}{n_2{}^2} \right) \frac{d^2f}{dt^2}$$

$$+ 8\pi^3 \left(2n_1{}^2 n_d + 2n_s{}^2 n_b \frac{n_1{}^2}{n_0{}^2} \right) \frac{df}{dt} + 16\pi^4 n_1{}^2 n_s{}^2 f(t)$$

$$= \frac{8\pi^2}{M_G + 2M_U} \left[(n_s{}^2 - N^2) P \sin 2\pi Nt + 2Nn_d P \cos 2\pi Nt \right].$$

If D_P denotes the central deflection due to a steady central load P,

$$D_P = \frac{P}{2\pi^2 n_0{}^2 M_G} = \frac{P}{2\pi^2 n_1{}^2 (M_G + 2M_U)},$$

and the differential equation for $f(t)$ takes the form

$$\frac{d^4f}{dt^4} + 2\pi \left(2n_b \frac{n_1{}^2}{n_0{}^2} + 2n_d \frac{n_1{}^2}{n_2{}^2} \right) \frac{d^3f}{dt^3} + 4\pi^2 \left(n_1{}^2 + 4n_b n_d \frac{n_1{}^2}{n_0{}^2} + n_s{}^2 \frac{n_1{}^2}{n_2{}^2} \right) \frac{d^2f}{dt^2}$$

$$+ 8\pi^3 \left(2n_1{}^2 n_d + 2n_s{}^2 n_b \frac{n_1{}^2}{n_0{}^2} \right) \frac{df}{dt} + 16\pi^4 n_1{}^2 n_s{}^2 f(t)$$

$$= 16\pi^4 \left[-n_1{}^2 (N^2 - n_s{}^2) \sin 2\pi Nt + 2Nn_d n_1{}^2 \cos 2\pi Nt \right] D_P.$$

The particular integral of this equation which gives the value of $f(t)$, after the oscillations have become steady, is

$$f(t) = D_P \frac{-(N^2 - n_s{}^2) \sin (2\pi Nt - \alpha) + 2Nn_d \cos (2\pi Nt - \alpha)}{\left\{ \left[\frac{N^4}{n_1{}^2} - N^2 \left(1 + \frac{4n_b n_d}{n_0{}^2} + \frac{n_s{}^2}{n_2{}^2} \right) + n_s{}^2 \right]^2 + \left[2N \left(n_d \frac{n_2{}^2 - N^2}{n_2{}^2} + n_b \frac{n_s{}^2 - N^2}{n_0{}^2} \right) \right]^2 \right\}^{\frac{1}{2}}}.$$

Write this

$$f(t) = D_P \frac{-(N^2 - n_s{}^2) \sin (2\pi Nt - \alpha) + 2Nn_d \cos (2\pi Nt - \alpha)}{\{ [\Phi(N)]^2 + [\Psi(N)]^2 \}^{\frac{1}{2}}}.$$

Then

$$\sin \alpha = \frac{\Psi(N)}{\{ [\Phi(N)]^2 + [\Psi(N)]^2 \}^{\frac{1}{2}}} \quad \text{and} \quad \cos \alpha = \frac{\Phi(N)}{\{ [\Phi(N)]^2 + [\Psi(N)]^2 \}^{\frac{1}{2}}},$$

and the maximum central deflection is

$$D_P \left\{ \frac{(N^2 - n_s{}^2)^2 + (2Nn_d)^2}{[\Phi(N)]^2 + [\Psi(N)]^2} \right\}^{\frac{1}{2}},$$

where
$$\Phi(N) = \frac{N^4}{n_1{}^2} - N^2\left(1 + \frac{4n_b n_d}{n_0{}^2} + \frac{n_s{}^2}{n_2{}^2}\right) + n_s{}^2$$

and
$$\Psi(N) = 2N\left(n_d\,\frac{n_2{}^2 - N^2}{n_2{}^2} + n_b\,\frac{n_s{}^2 - N^2}{n_0{}^2}\right).$$

The non-dimensional factor

$$\left\{\frac{(N^2 - n_s{}^2)^2 + (2Nn_d)^2}{[\Phi(N)]^2 + [\Psi(N)]^2}\right\}^{\frac{1}{2}}$$

may be termed the "Dynamic Magnifier", in that it gives the amount by which the steady central deflection due to a steady central force P is magnified when the force alternates N times per second.

For two similar locomotives situated at equal distances $\dfrac{d}{2}$ on either side of the centre, skidding their wheels so that their hammer-blows are in unison and of equal magnitude $P \sin 2\pi N t$, it can be shewn in a similar manner that the central dynamic deflection has the magnitude

$$2D_P \cos\frac{\pi d}{2l}\left\{\frac{(N^2 - n_s{}^2)^2 + (2Nn_d)^2}{[\Phi(N)]^2 + [\Psi(N)]^2}\right\}^{\frac{1}{2}},$$

where the expressions for $\Phi(N)$ and $\Psi(N)$ stated above still apply, but in this case

$$n_1{}^2 = n_0{}^2\,\frac{M_G}{M_G + 4M_U \cos^2\dfrac{\pi d}{2l}}, \qquad n_2{}^2 = n_0{}^2\,\frac{M_G}{M_G + 4(M_U + M_S)\cos^2\dfrac{\pi d}{2l}}.$$

D_P in this case is, again, the steady central deflection due to a load P statically applied at the centre of the bridge.

———————

The manner in which the dynamic magnifier varies with N is mainly decided by the variations in $N^2 - n_s{}^2$ and $\Phi(N)$.

When $N = n_s$, bridge oscillations are small, and this is only what might be expected, since for this particular frequency the oscillations of the locomotive on its spring have the vigour induced by resonance and, when the oscillations have become steady, the energy put into the system is mainly dissipated in overcoming spring friction, only a small residue being available for maintaining bridge oscillations.

When $\Phi(N) = 0$, the dynamic magnifier exhibits its peak values, and the condition for this is

$$N^4 - N^2 n_1{}^2\left(1 + \frac{4n_b n_d}{n_0{}^2} + \frac{n_s{}^2}{n_2{}^2}\right) + n_1{}^2 n_s{}^2 = 0.$$

This quadratic for N^2 will have real positive roots if

$$n_1 > \frac{2n_s}{1 + \dfrac{4n_b n_d}{n_0{}^2} + \dfrac{n_s{}^2}{n_2{}^2}}.$$

For bridges with natural frequencies of 5 or more this condition will be satisfied, and for such medium-span bridges the phenomenon of two critical frequencies will exist.

For practical purposes, the lower of these (say $N = N_1$) is of little interest, since, as will appear in a subsequent numerical example, it is only apparent as an "arrest" point in the rise of the deflection-frequency curve. For the higher critical frequency (say $N = N_2$) a very pronounced peak is developed and, in predicting the maximum state of oscillation in bridges of medium span, this higher critical frequency, provided that it lies within the range of practicable locomotive speeds, is the frequency which must be employed.

Without appreciable loss of accuracy the equation for determining the critical frequencies can be written

$$N^4 - N^2 n_1{}^2 \left(1 + \frac{n_s{}^2}{n_2{}^2}\right) + n_1{}^2 n_s{}^2 = 0,$$

and this important calculation can be effected without a knowledge of the two damping coefficients n_b and n_d, a fortunate circumstance, because, in the present state of experimental knowledge, it may be difficult to assign very precise numerical values to these two somewhat elusive characteristics.

To this degree of accuracy N_2, the important higher critical frequency, is given by

$$N_2{}^2 = n_1{}^2 \left[\frac{n_2{}^2 + n_s{}^2}{2n_2{}^2} + \sqrt{\left(\frac{n_2{}^2 + n_s{}^2}{2n_2{}^2}\right)^2 - \frac{n_s{}^2}{n_1{}^2}} \right],$$

and this is the fundamental frequency of the bridge when making free oscillations in the absence of bridge damping or spring friction.

For the extreme state of oscillation corresponding to $N = N_2$, the semi-amplitude of the central oscillation of the bridge is given by the comparatively simple formula

$$D_P \frac{\{(N_2{}^2 - n_s{}^2)^2 + (2N_2 n_d)^2\}^{\frac{1}{2}}}{2N_2 \left(n_d \dfrac{n_2{}^2 - N_2{}^2}{n_2{}^2} + n_b \dfrac{n_s{}^2 - N_2{}^2}{n_0{}^2} \right)}.$$

This method of computing maximum bridge oscillations will now be illustrated by deducing the deflection-frequency curve for a single-track bridge of about 150 feet span.

Bridge data.

$M_G = 300$ tons.

$n_0 = 6$.

$n_b = 0.36$.

This value of n_b gives a common ratio for successive residual deflections of about 0·68, a ratio which, although somewhat small for a span of 150 feet, is not an impossibility.

Locomotive data.

$M_S = 80$ tons.

$M_U = 20$ tons.

$P = 0.6N^2$ tons.

$n_s = 3$

$n_d = 0.50$.

This value of n_d gives a common ratio for successive residual spring oscillations of 0·35.

From these data it follows that

$$n_1{}^2 = n_0{}^2 \frac{M_G}{M_G + 2M_U} = 31.76 \quad \text{and} \quad n_2{}^2 = n_0{}^2 \frac{M_G}{M_G + 2M_U + 2M_S} = 21.60.$$

The critical frequencies are given by the equation

$$N^4 - N^2 n_1{}^2 \left(1 + \frac{4n_b n_d}{n_0{}^2} + \frac{n_s{}^2}{n_2{}^2} \right) + n_1{}^2 n_s{}^2 = 0,$$

i.e. $\qquad\qquad N^4 - 45.6354 N^2 + 285.8823 = 0.$

Hence $\qquad\qquad N^2 = 7.50 \quad \text{and} \quad 38.14,$

and $\qquad\qquad N_1 = 2.74, \quad N_2 = 6.18.$

If the deflection is expressed in the form $\dfrac{K}{100} D_W$, where D_W is the statical central deflection due to the total weight of the locomotive (100 tons) standing at the centre of the bridge,

$$K = \left\{ \frac{(N^2 - n_s{}^2)^2 + (2Nn_d)^2}{[\Phi(N)]^2 + [\Psi(N)]^2} \right\}^{\frac{1}{2}} \times 0.6N^2.$$

The values of $N^2 - n_s{}^2$, $2Nn_d$, $\Phi(N)$, $\Psi(N)$ and K, for values of N ranging from the limits $N = 0$ to $N = 10$, are tabulated below and the deflection-frequency curve obtained by plotting K in terms of N is shewn in Fig. 30.

N	$N^2 - n_s^2$	$2Nn_d$	$\Phi(N)$	$\Psi(N)$	K
0	-9	0	9·0000	0·0000	0·00
1	-8	1	7·5948	1·1137	0·63
2	-5	2	3·7570	1·8296	3·10
2·5	$-2·75$	2·5	1·2506	1·9141	6·04
3	0	3	$-1·3800$	1·7500	7·29
3·5	3·25	3·5	$-3·8750$	1·2876	8·60
4	7	4	$-5·9274$	0·4771	13·06
5	16	5	$-7·2407$	$-2·3870$	33·00
5·5	21·25	5·5	$-5·6516$	$-4·5400$	54·99
6	27	6	$-1·9200$	$-7·2399$	79·92
6·18	29·19	6·18	0·0000	$-8·3554$	81·81
6·5	33·25	6·5	4·4973	$-10·5365$	75·04
7	40	7	14·1841	$-14·4795$	59·80
8	55	8	46·0015	$-24·5036$	40·70
9	72	9	99·1801	$-37·7098$	33·53
10	91	10	180·1483	$-54·4960$	29·40

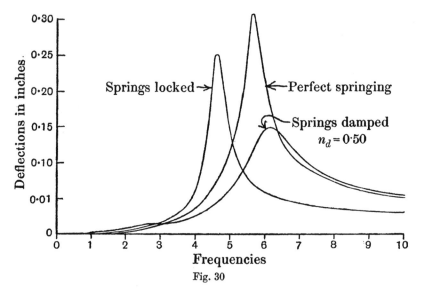

Fig. 30

From the formula $\qquad D_W = \dfrac{(M_U + M_S)g}{2\pi^2 n_0^2 M_G}$ feet,

$$D_W = 0·181 \text{ in.}$$

It will be seen that, in the neighbourhood of the lower critical frequency $N = 2·74$, the rise in the curve is checked, and that K attains its maximum value of about 82 in the vicinity of the higher critical frequency $N = 6·18$.

It is of interest to compare this figure with an actual deflection-frequency curve, see Fig. 31. This record, which is taken from the Report of the Bridge

Stress Committee, relates to Langport Bridge, a bridge of 112 feet span with
an unloaded frequency of 6·5 periods per second. The observations can be
seen to range themselves along a curve which has the same general character
as that represented by Fig. 30. The deflection first of all increases with the
frequency. This growth then receives a check, but at yet higher frequencies
a more vigorous increase is indicated leading up to a high frequency condition of resonance, though practical limitations of engine speed prevented
this peak value being surpassed or even fully attained.

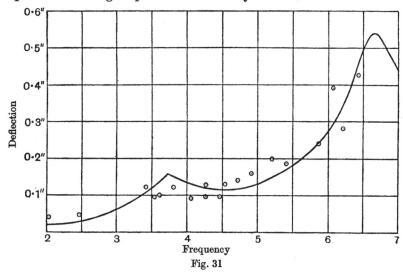

Fig. 31

The case of the locomotive with locked springs can be studied by making
n_d very large in the general expression for the dynamic magnifier, and the
value of the dynamic magnification thus obtained is

$$\frac{1}{\left\{\left(1-\frac{N^2}{n_2{}^2}\right)^2+\left(\frac{2Nn_b}{n_0{}^2}\right)^2\right\}^{\frac{1}{2}}}.$$

For comparison, the deflection-frequency curve for the case of the
locomotive with its springs locked is shewn on Fig. 30. For this case the
maximum deflection occurs when $N=N_2=4·65$, and is about 1·7 times the
maximum deflection for the case when spring movement was induced. A
comparison of these two curves shews the pronounced effect spring movement
has in reducing the magnitude of the maximum deflection due to hammer-
blow, and in altering the frequency at which this maximum is developed.

The accuracy of the method of predicting bridge oscillations for a moving
locomotive by means of a stationary bridge oscillator rests upon the validity
of the assumption that the oscillations mount up approximately to their

maximum in the brief period of time during which the locomotive is near the centre of the bridge, and this rate of rise in its turn depends upon the rapidity with which free oscillations subside into insignificance under the influence of the damping forces in the bridge and spring movement.

This rate of subsidence will now be investigated.

FREE OSCILLATIONS OF A BRIDGE SUPPORTING A SPRING-BORNE LOCOMOTIVE AT ITS CENTRE

The state of oscillation is given by $y = f(t) \sin \dfrac{\pi x}{l}$, where

$$\frac{d^4 f}{dt^4} + 2\pi \left(2n_b \frac{n_1^2}{n_0^2} + 2n_d \frac{n_1^2}{n_2^2}\right)\frac{d^3 f}{dt^3} + 4\pi^2 \left(n_1^2 + 4n_b n_d \frac{n_1^2}{n_0^2} + n_s^2 \frac{n_1^2}{n_2^2}\right)\frac{d^2 f}{dt^2}$$

$$+ 8\pi^3 \left(2n_1^2 n_d + 2n_s^2 n_b \frac{n_1^2}{n_0^2}\right)\frac{df}{dt} + 16\pi^4 n_1^2 n_s^2 f(t) = 0,$$

and is obtained by omitting the hammer-blow in the general equation formulated in the earlier part of this Chapter.

For a solution assume $f(t) = A e^{-2\pi \alpha t}\sin 2\pi \beta t.$

Substituting this value in the equation and equating to zero the coefficients of $e^{-2\pi \alpha t}\sin 2\pi \beta t$ and $e^{-2\pi \alpha t}\cos 2\pi \beta t$, the following two conditions for determining α and β are obtained:

$$(\alpha^4 + \beta^4 - 6\alpha^2\beta^2) + \left(2n_b \frac{n_1^2}{n_0^2} + 2n_d \frac{n_1^2}{n_2^2}\right)(3\alpha\beta^2 - \alpha^3)$$

$$+ \left(n_1^2 + 4n_b n_d \frac{n_1^2}{n_0^2} + n_s^2 \frac{n_1^2}{n_2^2}\right)(\alpha^2 - \beta^2) - \left(2n_1^2 n_d + 2n_s^2 n_b \frac{n_1^2}{n_0^2}\right)\alpha + n_1^2 n_s^2 = 0$$

and $$(4\alpha\beta^3 - 4\beta\alpha^3) + \left(2n_b \frac{n_1^2}{n_0^2} + 2n_d \frac{n_1^2}{n_2^2}\right)(3\alpha^2\beta - \beta^3)$$

$$- \left(n_1^2 + 4n_b n_d \frac{n_1^2}{n_0^2} + n_s^2 \frac{n_1^2}{n_2^2}\right)2\alpha\beta + \left(2n_1^2 n_d + 2n_s^2 n_b \frac{n_1^2}{n_0^2}\right)\beta = 0.$$

From these two equations, values of α and β are theoretically obtainable, but the general results are too complicated for practical purposes. Excellent approximations to the values of α and β can, however, be obtained by taking advantage of the fact that α is small in comparison with β.

On this assumption the first equation takes the form

$$\frac{\beta^4}{n_1^2} - \beta^2 \left(1 + \frac{4n_b n_d}{n_0^2} + \frac{n_s^2}{n_2^2}\right) + n_s^2 = 0,$$

giving $\beta = N_1$ and N_2, where N_1 and N_2 are the two critical frequencies found in connection with the oscillations produced by hammer-blow.

On this same assumption that α is small compared with β, the second equation takes the simplified form

$$\alpha\left[\frac{2\beta^2}{n_1^2}-\left(1+\frac{4n_b n_d}{n_0^2}+\frac{n_s^2}{n_2^2}\right)\right]-\left(\frac{n_b}{n_0^2}+\frac{n_d}{n_2^2}\right)\beta^2+\left(n_d+n_b\frac{n_s^2}{n_0^2}\right)=0,$$

and since

$$1+\frac{4n_b n_d}{n_0^2}+\frac{n_s^2}{n_2^2}=\frac{\beta^2}{n_1^2}+\frac{n_s^2}{\beta^2},$$

$$\alpha=\frac{n_b\dfrac{\beta^2-n_s^2}{n_0^2}+n_d\dfrac{\beta^2-n_2^2}{n_2^2}}{\dfrac{\beta^2}{n_1^2}-\dfrac{n_s^2}{\beta^2}}.$$

Hence fundamental free oscillations of the bridge have the general form

$$y=[e^{-2\pi\alpha_1 t}(A_1\sin 2\pi N_1 t+B_1\cos 2\pi N_1 t)$$
$$+e^{-2\pi\alpha_2 t}(A_2\sin 2\pi N_2 t+B_2\cos 2\pi N_2 t)]\sin\frac{\pi x}{l},$$

where N_1^2 and N_2^2 are the roots of the equation

$$\frac{N^4}{n_1^2}-N^2\left(1+\frac{4n_b n_d}{n_0^2}+\frac{n_s^2}{n_2^2}\right)+n_s^2=0$$

and

$$\alpha_1=\frac{n_b\dfrac{N_1^2-n_s^2}{n_0^2}+n_d\dfrac{N_1^2-n_2^2}{n_2^2}}{\dfrac{N_1^2}{n_1^2}-\dfrac{n_s^2}{N_1^2}},$$

$$\alpha_2=\frac{n_b\dfrac{N_2^2-n_s^2}{n_0^2}+n_d\dfrac{N_2^2-n_2^2}{n_2^2}}{\dfrac{N_2^2}{n_1^2}-\dfrac{n_s^2}{N_2^2}}.$$

For the numerical example previously considered in which $n_b=0\cdot36$, $n_d=0\cdot50$, $n_0^2=36$, $n_1^2=31\cdot76$, $n_2^2=21\cdot6$, $n_s^2=9$, the formulae above give

$$N_1=2\cdot74,\quad N_2=6\cdot18,\quad \alpha_1=0\cdot3542,\quad \alpha_2=0\cdot6995.$$

Using these results, it will now be shewn that the bridge oscillations due to a locomotive skidding its wheels at the centre of the span mount up to approximately their extreme value in a very short space of time.

Taking the case when the oscillations are most violent, that is, when the wheel revolutions are N_2 per second, the central deflection h, after the disturbance has become steady, may be written

$$h=D\sin 2\pi N_2 t.$$

The corresponding displacement of the spring-borne mass can be deduced from this by the equation

$$\frac{d^2z}{dt^2} + 4\pi n_d \frac{dz}{dt} + 4\pi^2 n_s^2 z = 4\pi^2 n_s^2 h + 4\pi n_d \frac{dh}{dt},$$

and this gives for the steady state of oscillation of the spring-borne mass

$$z = -D\left[0.252 \sin 2\pi N_2 t + 0.265 \cos 2\pi N_2 t\right].$$

For the free oscillation which has to be introduced to satisfy the starting condition, let

$$h = e^{-2\pi\alpha_1 t}(A_1 \sin 2\pi N_1 t + B_1 \cos 2\pi N_1 t)$$
$$+ e^{-2\pi\alpha_2 t}(A_2 \sin 2\pi N_2 t + B_2 \cos 2\pi N_2 t).$$

The corresponding oscillation induced in the spring-borne mass is given by

$$z = e^{-2\pi\alpha_1 t}[(5.87A_1 + 1.54B_1)\sin 2\pi N_1 t + (-1.5A_1 + 5.87B_1)\cos 2\pi N_1 t]$$
$$+ e^{-2\pi\alpha_2 t}[(-0.30A_2 + 0.19B_2)\sin 2\pi N_2 t + (-0.19A_2 - 0.30B_2)\cos 2\pi N_2 t].$$

Combining the steady forced oscillations with the free oscillations to satisfy the four conditions, that both girder and spring-borne mass start from rest and from a position of statical equilibrium, the following relations are obtained:

Since $h = 0$ when $t = 0$, $\qquad B_1 + B_2 = 0.$

Since $z = 0$ when $t = 0$,

$$-1.54A_1 - 0.19A_2 + 6.17B_1 = 0.265D.$$

Since $\dfrac{dh}{dt} = 0$ when $t = 0$,

$$2.74A_1 + 6.18A_2 + 0.35B_1 = -6.183D.$$

Since $\dfrac{dz}{dt} = 0$ when $t = 0$,

$$16.63A_1 - 1.72A_2 + 0.74B_1 = 1.557D.$$

Hence $A_2 = -0.996D$, $A_1 = -0.003D$, $B_1 = 0.002D$, $B_2 = -0.002D$,

and, to a high degree of accuracy, the way in which the oscillation of the bridge mounts up from a condition of rest is given by

$$h = D[1 - e^{-2\pi\alpha_2 t}]\sin 2\pi N_2 t.$$

After only half a second the semi-amplitude of h is $D[1 - e^{-\pi\alpha_2}]$, that is $D[1 - e^{-2.1975}] = 0.89D$, and in one second the oscillation is 99 per cent. of its full value. From this case it appears that if the moving locomotive can get in three hammer-blows while it is in the central region of the bridge, the oscillation set up will probably be about 90 per cent. of that induced if it remained hammering at the centre of the bridge for an indefinite length of time, and this calculation confirms the belief that bridge oscillations determined with reference to a stationary skidding locomotive are not likely, in the case of medium-span bridges, to be much in excess of those generated by a locomotive in motion.

This consideration, however, is one of such fundamental importance that it calls for the determination of the oscillations due to the moving locomotive by the fullest possible mathematical analysis. In this way, laborious though the work may be, a standard can be set up whereby the degree of accuracy achieved by approximate methods becomes apparent.

FULL ANALYTICAL TREATMENT OF THE OSCILLATIONS PRODUCED BY A SPRING-BORNE LOCOMOTIVE MOVING ACROSS A BRIDGE

Adopting the notation defined in an earlier part of this Chapter, let h and z be the depths of M_U and M_S below their mean positions at any instant. The pressure exerted between the wheels and the rails is

$$P \sin 2\pi Nt - M_U \frac{d^2h}{dt^2} - M_S \frac{d^2z}{dt^2},$$

and the equivalent primary harmonic distribution for this concentrated load at any instant is

$$\frac{2}{l}\left[P \sin 2\pi Nt - M_U \frac{d^2h}{dt^2} - M_S \frac{d^2z}{dt^2} \right] \sin 2\pi nt \sin \frac{\pi x}{l},$$

where $n = \dfrac{v}{2l}$.

The equation of motion of the bridge is accordingly

$$EI \frac{d^4y}{dx^4} + 4\pi n_b m \frac{dy}{dt} + m \frac{d^2y}{dt^2}$$

$$= \frac{2}{l}\left[P \sin 2\pi Nt - M_U \frac{d^2h}{dt^2} - M_S \frac{d^2z}{dt^2} \right] \sin 2\pi nt \sin \frac{\pi x}{l}$$

$$\dots\dots(3).$$

For the solution take $\qquad y = f(t) \sin \dfrac{\pi x}{l}.$

Consequently $\qquad\qquad h = f(t) \sin 2\pi nt,$

$$\frac{dh}{dt} = \frac{df}{dt} \sin 2\pi nt + 2\pi n f(t) \cos 2\pi nt$$

and $\qquad \dfrac{d^2h}{dt^2} = \dfrac{d^2f}{dt^2} \sin 2\pi nt + 4\pi n \dfrac{df}{dt} \cos 2\pi nt - 4\pi^2 n^2 f(t) \sin 2\pi nt.$

Substituting these values for y and h in equation (3), an equation for $f(t)$ is obtained in the form

$$4\pi^2 n_0{}^2 M_G f(t) + 4\pi n_b M_G \frac{df}{dt} + M_G \frac{d^2 f}{dt^2}$$

$$= 2\left[P \sin 2\pi N t - M_U \sin 2\pi n t \frac{d^2 f}{dt^2} - 4\pi n M_U \cos 2\pi n t \frac{df}{dt} \right.$$

$$\left. + 4\pi^2 n^2 M_U \sin 2\pi n t\, f(t) - M_S \frac{d^2 z}{dt^2} \right] \sin 2\pi n t,$$

that is,

$$\frac{d^2 f}{dt^2} (M_G + M_U - M_U \cos 4\pi n t) + 2\pi \frac{df}{dt} (2n_b M_G + 2n M_U \sin 4\pi n t)$$

$$+ 4\pi^2 f(t)(n_0{}^2 M_G - n^2 M_U + n^2 M_U \cos 4\pi n t)$$

$$= P[\cos 2\pi (N-n)t - \cos 2\pi (N+n)t] - 2M_S \frac{d^2 z}{dt^2} \sin 2\pi n t \ldots\ldots(4).$$

The equation of motion of M_S as previously established in this Chapter is

$$\frac{d^2 z}{dt^2} + 4\pi n_d \frac{dz}{dt} + 4\pi^2 n_s{}^2 z = 4\pi^2 n_s{}^2 h + 4\pi n_d \frac{dh}{dt},$$

that is,

$$\frac{d^2 z}{dt^2} + 4\pi n_d \frac{dz}{dt} + 4\pi^2 n_s{}^2 z$$

$$= 4\pi^2 [n_s{}^2 \sin 2\pi n t + 2n n_d \cos 2\pi n t] f(t) + 2\pi [2n_d \sin 2\pi n t] \frac{df}{dt}$$

$$\ldots\ldots(5).$$

Between these equations (4) and (5), it is possible to eliminate z and obtain a differential equation for $f(t)$. This equation will be of the fourth order, and since its coefficients are functions of t, a solution can only be achieved in the form of an infinite series of the character employed in the analysis of the long-span bridge oscillations.

For the particular integral or forced oscillation the form of the series is

$$f(t) = \Sigma\, (A_r \cos 2\pi q_r t + B_r \sin 2\pi q_r t),$$

where $q_r = N + rn$, and r is any odd integer positive or negative.

On substituting this series value of $f(t)$ in equation (4), it will be found that the coefficient of $\cos 2\pi q_r t$, on the left-hand side of the equation, is

$$4\pi^2 \{ A_r [(n_0{}^2 - q_r{}^2) M_G - (n^2 + q_r{}^2) M_U] + A_{r+2} [\tfrac{1}{2} q^2_{r+1} M_U]$$

$$+ A_{r-2} [\tfrac{1}{2} q^2_{r-1} M_U] + B_r [2n_b q_r M_G] \},$$

and the coefficient of $\sin 2\pi q_r t$ is

$$4\pi^2 \{ B_r [(n_0{}^2 - q_r{}^2) M_G - (n^2 + q_r{}^2) M_U] + B_{r+2} [\tfrac{1}{2} q^2_{r+1} M_U]$$

$$+ B_{r-2} [\tfrac{1}{2} q^2_{r-1} M_U] - A_r [2n_b q_r M_G] \}.$$

The next step is the determination of the coefficients of $\cos 2\pi q_r t$ and $\sin 2\pi q_r t$ in the term $-2M_s \dfrac{d^2z}{dt^2} \sin 2\pi nt$, which appears on the right-hand side of equation (4).

Putting $\qquad\qquad f(t) = A_r \cos 2\pi q_r t + B_r \sin 2\pi q_r t$

in equation (5), the value of z obtained can be written in the form

$$- (\beta_{r+1} A_r + \alpha_{r+1} B_r) \cos 2\pi q_{r+1} t + (\beta_{r-1} A_r + \alpha_{r-1} B_r) \cos 2\pi q_{r-1} t$$
$$+ (\alpha_{r+1} A_r - \beta_{r+1} B_r) \sin 2\pi q_{r+1} t - (\alpha_{r-1} A_r - \beta_{r-1} B_r) \sin 2\pi q_{r-1} t,$$

where α_r denotes $\qquad \dfrac{1}{2} \left[\dfrac{n_s^2 (n_s^2 - q_r^2) + (2n_d q_r)^2}{(n_s^2 - q_r^2)^2 + (2n_d q_r)^2} \right] q_r^2$

and β_r denotes $\qquad \left[\dfrac{n_d q_r^3}{(n_s^2 - q_r^2)^2 + (2n_d q_r)^2} \right] q_r^2.$

From this it follows that in the general expression for $-2M_s \dfrac{d^2z}{dt^2} \sin 2\pi nt$, the coefficient of $\sin 2\pi q_r t$ is

$$4\pi^2 M_s [(\beta_{r+1} + \beta_{r-1}) A_r - \beta_{r+1} A_{r+2} - \beta_{r-1} A_{r-2}$$
$$+ (\alpha_{r+1} + \alpha_{r-1}) B_r - \alpha_{r+1} B_{r+2} - \alpha_{r-1} B_{r-2}],$$

and the coefficient of $\cos 2\pi q_r t$ is

$$4\pi^2 M_s [(\alpha_{r+1} + \alpha_{r-1}) A_r - \alpha_{r+1} A_{r+2} - \alpha_{r-1} A_{r-2}$$
$$- (\beta_{r+1} + \beta_{r-1}) B_r + \beta_{r+1} B_{r+2} + \beta_{r-1} B_{r-2}].$$

Let $\qquad\qquad (n_0^2 - q_r^2) \dfrac{M_G}{M_S} - (n^2 + q_r^2) \dfrac{M_U}{M_S} = a_r,$

$$\tfrac{1}{2} q_r^2 \dfrac{M_U}{M_S} = b_r$$

and $\qquad\qquad 2 n_b q_r \dfrac{M_G}{M_S} = c_r.$

By equating coefficients of $\cos 2\pi q_r t$ and $\sin 2\pi q_r t$ on the two sides of equation (4), the condition that

$$f(t) = \Sigma (A_r \cos 2\pi q_r t + B_r \sin 2\pi q_r t)$$

fits the equation is

$$(a_r - \alpha_{r+1} - \alpha_{r-1}) A_r + (b_{r+1} + \alpha_{r+1}) A_{r+2} + (b_{r-1} + \alpha_{r-1}) A_{r-2}$$
$$+ (c_r + \beta_{r+1} + \beta_{r-1}) B_r - \beta_{r+1} B_{r+2} - \beta_{r-1} B_{r-2} = 0$$

for all values of r except when $r = -1$ and $r = +1$.

When $r = -1$, the expression has the value

$$\frac{P}{4\pi^2 M_S} = \frac{M_G}{2M_S} n_0^2 D_P,$$

and when $r = +1$, it has the value

$$-\frac{M_G}{2M_S} n_0^2 D_P.$$

By equating the coefficients of $\sin 2\pi q_r t$ on the two sides of equation (4),

$$(a_r - \alpha_{r+1} - \alpha_{r-1}) B_r + (b_{r+1} + \alpha_{r+1}) B_{r+2} + (b_{r-1} + \alpha_{r-1}) B_{r-2}$$
$$- (c_r + \beta_{r+1} + \beta_{r-1}) A_r + \beta_{r+1} A_{r+2} + \beta_{r-1} A_{r-2} = 0$$

for all values of r.

From these conditions, a series of equations for determining the A and B coefficients can be laid down.

When this lengthy and laborious process has been accomplished the connection between δ, the central deflection, and a, the distance the locomotive has moved along the bridge, can be expressed in the form

$$\delta = \Phi(a) \cos 2\pi N t + \Psi(a) \sin 2\pi N t,$$

where

$$\Phi(a) = \Sigma \left[(A_r + A_{-r}) \cos \frac{\pi r a}{l} + (B_r - B_{-r}) \sin \frac{\pi r a}{l} \right]$$

and

$$\Psi(a) = \Sigma \left[(A_{-r} - A_r) \sin \frac{\pi r a}{l} + (B_r + B_{-r}) \cos \frac{\pi r a}{l} \right].$$

To satisfy the starting conditions a state of free oscillation must be added, that is, a solution of equations (4) and (5) with P made zero. For practical purposes this addition is of little interest, since, owing to the heavy damping introduced by spring friction, this free oscillation dies out very rapidly and is certainly negligible by the time the locomotive has reached the centre of the bridge.

The foregoing analysis will now be applied to the case previously considered in which

$$M_G = 300 \text{ tons}, \quad l = 150 \text{ ft.}, \quad n_0 = 6, \quad n_b = 0.36,$$

$$M_S = 80 \text{ tons}, \quad M_U = 20 \text{ tons}, \quad n_s = 3, \quad n_d = 0.50, \quad P = 0.6N^2 \text{ tons},$$

and the circumference of the driving wheels is 15 feet.

The value taken for $N = 6.18$, since previous considerations have shewn that this is the particular frequency which will give rise to the most violent state of bridge oscillation.

A preliminary survey shewed that a very accurate estimate of the oscillation could be obtained by limiting the series

$$f(t) = \Sigma \left[A_r \cos 2\pi q_r t + B_r \sin 2\pi q_r t \right]$$

within the range $r = -7$ to $r = +7$.

The numerical values of q_r, a_r, b_r, c_r, α_r, β_r for the required range of values of r are given in the following table:

r	q_r	a_r	b_r	c_r	α_r	β_r
8	8·652	− 164·4523	9·3571	23·3604	− 4·3932	5·4943
7	8·343	− 143·4463	8·7007	22·5261	− 4·4249	5·4001
6	8·034	− 123·2047	8·0682	21·6918	− 4·4607	5·3131
5	7·725	− 103·7263	7·4595	20·8575	− 4·5012	5·2347
4	7·416	− 85·0123	6·8746	20·0023	− 4·5475	5·1667
3	7·107	− 67·0615	6·3137	19·1889	− 4·6005	5·1116
2	6·798	− 49·8751	5·7766	18·3546	− 4·6617	5·0726
1	6·489	− 33·4523	5·2634	17·5203	− 4·7327	5·0541
0	6·180	− 17·7935	4·7741	16·6860	− 4·8157	5·0621
−1	5·871	− 2·8983	4·3086	15·8517	− 4·9133	5·1054
−2	5·562	11·2329	3·8670	15·0174	− 5·0285	5·1971
−3	5·253	24·6001	3·4493	14·1831	− 5·1647	5·3569
−4	4·944	37·2037	3·0554	13·3488	− 5·3707	5·6172
−5	4·635	49·0433	2·6854	12·5145	− 5·5046	6·0322
−6	4·326	60·1189	2·3393	11·6802	− 5·6859	6·6990
−7	4·017	70·4309	2·0170	10·8459	− 5·7856	7·7983
−8	3·708	88·7633	1·7187	10·0116	− 5·4903	9·6539

The sixteen equations for determining the sixteen coefficients A_7, A_5, ... A_{-5}, A_{-7}; B_7, B_5, ... B_{-5}, B_{-7} can now be written down as follows, on the assumption that coefficients outside the range $r = +7$ to $r = -7$ are negligible:

(1) $-134 \cdot 5924 A_7 + 3 \cdot 6075 A_5 + 33 \cdot 3335 B_7 - 5 \cdot 3131 B_5 = 0,$

(2) $- 94 \cdot 7181 A_5 + 3 \cdot 6075 A_7 + 2 \cdot 3271 A_3 + 31 \cdot 3373 B_5$
$- 5 \cdot 3131 B_7 - 5 \cdot 1667 B_3 = 0,$

(3) $- 57 \cdot 8523 A_3 + 2 \cdot 3271 A_5 + 1 \cdot 1149 A_1 + 29 \cdot 4282 B_3$
$- 5 \cdot 1667 B_5 - 5 \cdot 0726 B_1 = 0,$

(4) $- 23 \cdot 9749 A_1 + 1 \cdot 1149 A_3 - 0 \cdot 0416 A_{-1} + 27 \cdot 6650 B_1$
$- 5 \cdot 0726 B_3 - 5 \cdot 0621 B_{-1} = - 15 \cdot 4679 D_W,$

(5) $+ 6 \cdot 9459 A_{-1} - 0 \cdot 0416 A_1 - 1 \cdot 1615 A_{-3} + 26 \cdot 1109 B_{-1}$
$- 5 \cdot 0621 B_1 - 5 \cdot 1971 B_{-3} = + 15 \cdot 4679 D_W,$

(6) $+ 34 \cdot 9993 A_{-3} - 1 \cdot 1615 A_{-1} - 2 \cdot 3153 A_{-5} + 24 \cdot 9974 B_{-3}$
$- 5 \cdot 1971 B_{-1} - 5 \cdot 6172 B_{-5} = 0,$

(7) $+ 60 \cdot 0999 A_{-5} - 2 \cdot 3153 A_{-3} - 3 \cdot 3466 A_{-7} + 24 \cdot 8307 B_{-5}$
$- 5 \cdot 6172 B_{-3} - 6 \cdot 6990 B_{-7} = 0,$

(8) $+ 81 \cdot 6071 A_{-7} - 3 \cdot 3466 A_{-5} + 27 \cdot 1988 B_{-7} - 6 \cdot 6990 B_{-5} = 0,$

$$(9) \quad -134\cdot5924B_7 \; +3\cdot6075B_5 \; -33\cdot3335A_7 \; + \; 5\cdot3131A_5 \qquad =0,$$

$$(10) \quad - \; 94\cdot7181B_5 \; +3\cdot6075B_7 \; + \; 2\cdot3271B_3 \; -31\cdot3373A_5$$
$$+5\cdot3131A_7 \; +5\cdot1667A_3 =0,$$

$$(11) \quad - \; 57\cdot8523B_3 \; +2\cdot3271B_5 \; + \; 1\cdot1149B_1 \; -29\cdot4282A_3$$
$$+5\cdot1667A_5 \; +5\cdot0726A_1 \; =0,$$

$$(12) \quad - \; 23\cdot9749B_1 \; +1\cdot1149B_3 \; - \; 0\cdot0416B_{-1} -27\cdot6650A_1$$
$$+5\cdot0726A_3 \; +5\cdot0621A_{-1}=0,$$

$$(13) \quad + \; 6\cdot9459B_{-1}-0\cdot0416B_1 \; - \; 1\cdot1615B_{-3}-26\cdot1109A_{-1}$$
$$+5\cdot0621A_1 \; +5\cdot1971A_{-3}=0,$$

$$(14) \quad + \; 34\cdot9993B_{-3}-1\cdot1615B_{-1}- \; 2\cdot3153B_{-5}-24\cdot9974A_{-3}$$
$$+5\cdot1971A_{-1}+5\cdot6172A_{-5}=0,$$

$$(15) \quad + \; 60\cdot0999B_{-5}-2\cdot3153B_{-3}- \; 3\cdot3466B_{-7}-24\cdot8307A_{-5}$$
$$+5\cdot6172A_{-3}+6\cdot6990A_{-7}=0,$$

$$(16) \quad + \; 81\cdot6071B_{-7}-3\cdot3466B_{-5}-27\cdot1988A_{-7}+ \; 6\cdot6990A_{-5} \qquad =0.$$

From these equations the following results are obtained:

$A_1=0\cdot25423D_W$	$A_{-1}=0\cdot19379D_W$	$B_1= -0\cdot24720D_W$	$B_{-1}= \;\;\; 0\cdot50187D_W$
$A_3=0\cdot02819D_W$	$A_{-3}=0\cdot05989D_W$	$B_3= \;\;\; 0\cdot00333D_W$	$B_{-3}= \;\;\; 0\cdot02957D_W$
$A_5=0\cdot00094D_W$	$A_{-5}=0\cdot00589D_W$	$B_5= \;\;\; 0\cdot00131D_W$	$B_{-5}= -0\cdot00208D_W$
$A_7=0\cdot00000D_W$	$A_{-7}=0\cdot00024D_W$	$B_7= \;\;\; 0\cdot00007D_W$	$B_{-7}= -0\cdot00048D_W$

Writing the central deflection in the form

$$\delta = [\Phi\,(a)\cos 2\pi Nt + \Psi\,(a)\sin 2\pi Nt]\,D_W,$$

where a is the distance the locomotive has moved along the bridge, the values of $\Phi\,(a)$ and $\Psi\,(a)$ for various values of a are shewn in the following table:

a	0	$\dfrac{l}{16}$	$\dfrac{l}{8}$	$\dfrac{3l}{16}$	$\dfrac{l}{4}$	$\dfrac{5l}{16}$	$\dfrac{3l}{8}$	$\dfrac{7l}{16}$	$\dfrac{l}{2}$
$\Phi\,(a)$	0·5431	0·3592	0·1372	−0·0932	−0·3012	−0·4670	−0·5865	−0·6672	−0·7201
$\Psi\,(a)$	0·2864	0·2869	0·2595	0·2039	0·1333	0·0600	−0·0029	−0·0516	−0·0872

a	l	$\dfrac{15l}{16}$	$\dfrac{7l}{8}$	$\dfrac{13l}{16}$	$\dfrac{3l}{4}$	$\dfrac{11l}{16}$	$\dfrac{5l}{8}$	$\dfrac{9l}{16}$	$\dfrac{l}{2}$
$\Phi\,(a)$	−0·5431	−0·6738	−0·7524	−0·7902	−0·8008	−0·7950	−0·7791	−0·7550	−0·7201
$\Psi\,(a)$	−0·2864	−0·2665	−0·2377	−0·2089	−0·1809	−0·1574	−0·1361	−0·1138	−0·0872

The central deflection thus obtained is only the forced oscillation and, to satisfy the conditions that the oscillation builds up from a state of zero deflection and zero velocity, a state of free oscillation has to be superposed. This adjustment need not be carried out with great precision because, since before the locomotive has reached the central region of the bridge, this state of free oscillation will have been damped into insignificance. The frequency of the free oscillations will vary from 6 at the start to 6·18 when the locomotive has reached the centre of the span, and in this distance the damping factor will increase from $e^{-2\cdot26t}$ to $e^{-4\cdot40t}$. Without more detailed examination this gives a sufficiently accurate estimate of the free oscillation.

The result of superposing the forced oscillation on the "crawl-deflection" and introducing a free oscillation to satisfy the starting condition is shewn by the upper graph of Fig. 32.

From this it appears that the dynamical increment of deflection when the locomotive is passing the centre of the bridge is $\dfrac{72\cdot5}{100}D_W$. The greatest state

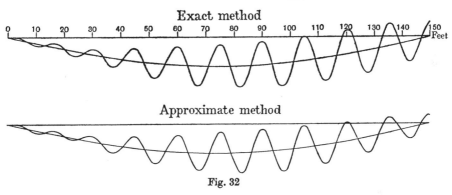

Fig. 32

of oscillation occurs when the locomotive has passed the centre of the bridge by a distance of about 40 feet, the semi-amplitude of the oscillation being $\dfrac{82\cdot1}{100}D_W$. The maximum dynamical increment of deflection is $\dfrac{75}{100}D_W$.

The main purpose of this lengthy calculation was to obtain a standard for ascertaining how far it is legitimate in calculating dynamical deflections to replace a moving locomotive by a stationary locomotive skidding its wheels at the centre of the bridge.

The dynamical increment of the central deflection deduced by the skidding locomotive method was $\dfrac{81\cdot8}{100}D_W$, and this is less than 10 per cent. in excess of the value $\dfrac{75}{100}D_W$ obtained by the full mathematical analysis.

Though the skidding locomotive method for predicting the maximum state of oscillations in bridges of medium span would appear to be sufficiently accurate for all practical purposes, nevertheless, if required, a yet closer approximation can be obtained by utilizing the principle previously established that in its deflecting effect a moving hammer-blow $P\sin2\pi Nt$ is equivalent to two central alternating forces

$$\frac{P}{2}\cos 2\pi(N-n)t \quad \text{and} \quad -\frac{P}{2}\cos 2\pi(N+n)t,$$

where $n=\dfrac{v}{2l}$.

If the locomotive is imagined to be standing at the centre of the bridge and its unsprung mass is subjected to these two alternating forces, the effect of a moving hammer-blow is achieved; but, as a method of determining bridge oscillations, this system is still defective in that it does not take full account of the variations in the inertia effect and spring damping as the locomotive moves along the bridge. Nevertheless, for the case of most practical importance, that is, when $N = N_2$ and resonance occurs, this double oscillator method gives a representation of the state of oscillation which is in remarkably close agreement with the results obtained by the full analytical treatment.

The analysis for the double oscillator is only a slight amplification of the analysis of the single oscillator given earlier on in this Chapter and will be merely outlined as follows.

If D_P is the central deflection produced by P when treated as a central statical force, the forced oscillation due to $\dfrac{P}{2}\cos 2\pi(N-n)t$ is

$$-\frac{D_P}{2}\frac{[(N-n)^2-n_s{}^2]\cos[2\pi(N-n)t-\alpha]+2(N-n)n_d\sin[2\pi(N-n)t-\alpha]}{\{[\Phi(N-n)]^2+[\Psi(N-n)]^2\}^{\frac{1}{2}}},$$

where $\Phi(N-n)$ and $\Psi(N-n)$ denote the values of Φ and Ψ when in the general expressions for these functions N is replaced by $N-n$ and

$$\sin\alpha=\frac{\Psi(N-n)}{\{[\Phi(N-n)]^2+[\Psi(N-n)]^2\}^{\frac{1}{2}}}, \quad \cos\alpha=\frac{\Phi(N-n)}{\{[\Phi(N-n)]^2+[\Psi(N-n)]^2\}^{\frac{1}{2}}}.$$

The contribution made by $-\dfrac{P}{2}\cos 2\pi(N+n)t$, can be written down in a similar form, $N-n$ being replaced by $N+n$.

Combined together, the central deflection δ can be expressed in the form

$$\delta=[A\cos 2\pi nt+B\sin 2\pi nt]\cos 2\pi Nt+[C\cos 2\pi nt+D\sin 2\pi nt]\sin 2\pi Nt,$$

and since $2\pi nt=\dfrac{\pi vt}{l}=\dfrac{\pi a}{l}$, where a is the distance the locomotive has advanced along the bridge, the connection between δ and a can be expressed in the form

$$\delta=\left[A\cos\frac{\pi a}{l}+B\sin\frac{\pi a}{l}\right]\cos 2\pi Nt+\left[C\cos\frac{\pi a}{l}+D\sin\frac{\pi a}{l}\right]\sin 2\pi Nt.$$

For the numerical example, which was worked out by the full mathematical analysis, $N = 6\cdot18$ and $n = 0\cdot31$. Hence

$$N + n = 6\cdot49, \quad N - n = 5\cdot87,$$

$$\Phi(N-n) = -3\cdot1258, \quad \Psi(N-n) = -6\cdot4826,$$

$$\Phi(N+n) = 4\cdot3387, \quad \Psi(N+n) = -10\cdot4644,$$

and the (δ, a) connection is given by

$$\delta = \frac{D_W}{100}\left\{\left(28\cdot044\cos\frac{\pi a}{l} - 69\cdot028\sin\frac{\pi a}{l}\right)\cos 2\pi Nt \right.$$
$$\left. + \left(12\cdot141\cos\frac{\pi a}{l} - 9\cdot725\sin\frac{\pi a}{l}\right)\sin 2\pi Nt\right\}.$$

This forced oscillation, superposed on the "crawl-deflection" and combined with a free oscillation to adjust the starting conditions, gives the state of oscillation depicted by the lower graph of Fig. 32, and it will be seen that it is in remarkably close agreement with the upper graph which was obtained by the full mathematical analysis. The greatest dynamical increment of deflection is $\frac{72\cdot2}{100} D_W$ and occurs when the locomotive has passed the centre of the bridge by about 10 feet, and this value differs by less than 4 per cent. from the value $\frac{75}{100} D_W$ obtained by the exact process.

The double oscillator method makes it possible to study the variations in the oscillations as a locomotive passes along a bridge, but when, as is usually the case, only the maximum state of oscillation is required, the single oscillator method owing to its comparative simplicity is to be preferred, and the small over-estimate which results from the application of this method can hardly be regarded as a disadvantage. Hence the important conclusion which emerges from this Chapter is that the maximum state of oscillation generated in a bridge of medium span occurs when the engine revolutions are N_2 per second, where $N_2{}^2$ is the larger of the two roots of the equation

$$N^4 - N^2 n_1{}^2\left(1 + \frac{n_s{}^2}{n_2{}^2}\right) + n_1{}^2 n_s{}^2 = 0,$$

and the corresponding dynamical increment of the central deflection is

$$D_P \frac{\{(N_2{}^2 - n_s{}^2)^2 + (2N_2 n_d)^2\}^{\frac{1}{2}}}{2N_2\left[n_d\dfrac{n_2{}^2 - N_2{}^2}{n_2{}^2} + n_b\dfrac{n_s{}^2 - N_2{}^2}{n_0{}^2}\right]},$$

where D_P is the central deflection for the hammer-blow at N_2 revs. per sec., treated as a central statical force.

CHAPTER X

RESISTANCE TO SPRING MOVEMENT TREATED IN ACCORDANCE WITH THE LAWS OF SOLID FRICTION

In the analysis developed in Chapter IX, it was assumed that the resistance to spring movements was of the fluid friction variety, the force of friction varying with the velocity of M_S relative to M_U. Actually the resistance to spring movement is more in the nature of dry, solid-body friction, in which the resistance to movement is independent of velocity.

The treatment of these two cases is not so different as might appear at first sight, and an extension of the analysis given in the preceding Chapter can be made to include the effects of spring movements damped by solid-body friction.

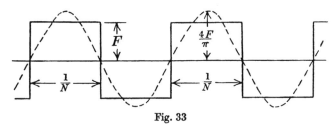

Fig. 33

As the springs oscillate N times per second the variation in the frictional resistance drawn to a time basis will be in accordance with the graph shewn in Fig. 33, F being the value of the frictional force which changes its direction periodically but always has the same magnitude.

This rectangular-shaped graph can be analysed into the harmonic series

$$\frac{4F}{\pi}\left[\sin 2\pi Nt + \tfrac{1}{3}\sin 6\pi Nt + \tfrac{1}{5}\sin 10\pi Nt + \ldots\right],$$

and, with the notation adopted in the preceding Chapter, the relative displacement $z-h$ of M_S relative to M_U, produced by h, an alternating displacement of M_U, is given by the equation

$$\frac{d^2(z-h)}{dt^2} + 4\pi^2 n_s^2(z-h)$$

$$= -\frac{d^2h}{dt^2} - \frac{4F}{\pi M_S}\left[\sin 2\pi Nt + \tfrac{1}{3}\sin 6\pi Nt + \tfrac{1}{5}\sin 10\pi Nt + \ldots\right].$$

The contribution to $(z-h)$ made by friction is accordingly

$$\frac{F}{\pi^3 M_s}\left[\frac{\sin 2\pi Nt}{N^2 - n_s^2} + \frac{1}{3}\frac{\sin 6\pi Nt}{9N^2 - n_s^2} + \frac{1}{5}\frac{\sin 10\pi Nt}{25N^2 - n_s^2} + \cdots\right].$$

In bridges of moderate span, oscillations of practical importance only occur when terms such as $3(9N^2 - n_s^2)$ and $5(25N^2 - n_s^2)$ are very large in comparison with $N^2 - n_s^2$. Consequently, the value of $z-h$ will be practically unaffected by any of the harmonic components of F other than the first, and no appreciable errors in calculating bridge oscillations will be introduced if the frictional force is replaced by its fundamental harmonic component, which is shewn by the dotted curve in Fig. 33.

From the point of view of energy expenditure no error is introduced by this approximation, since the work done per cycle by the alternating force $4F\sin 2\pi Nt$, and by the force F, varying in accordance with the rectangular graph, are the same for a given amplitude of $z-h$.

Replacing the frictional force by an harmonically varying resistance is in effect the same as replacing solid friction by an equivalent fluid friction. In the mathematics of the bridge oscillator, the frictional resistance to spring movement was taken to be $4\pi n_d M_s v$, where v is the velocity of M_s relative to M_U. This resistance varies harmonically in time, its crest value being $4\pi n_d M_s \times \pi Ne$, where e is the amplitude of the movement of M_s relative to M_U.

Hence by putting $4\pi^2 n_d M_s Ne = \dfrac{4F}{\pi}$, that is $n_d = \dfrac{F}{\pi^3 M_s Ne}$, the bridge oscillator methods are made applicable to the case of spring damping, which varies according to the laws of solid-body friction. In applying this process, an initial difficulty is encountered, arising from the fact that n_d cannot be evaluated until e is known, and conversely, the direct determination of e requires a knowledge of n_d. This difficulty can be circumvented as follows.

The state of oscillation of the bridge, due to the stationary skidding locomotive, can be expressed in the form $\delta = \delta_0 \sin 2\pi Nt$, where

$$\delta_0 = D_P\left\{\frac{(N^2 - n_s^2)^2 + (2Nn_d)^2}{[\Phi(N)]^2 + [\Psi(N)]^2}\right\}^{\frac{1}{2}}.$$

The frictional resistance, deduced from the expression

$$4\pi n_d M_s \frac{d}{dt}(z-\delta),$$

has for its crest value

$$4\pi n_d M_s \frac{2\pi N^3 \delta_0}{\{(N^2 - n_s^2)^2 + (2Nn_d)^2\}^{\frac{1}{2}}} = \frac{8\pi^2 N^3 n_d M_s}{\{[\Phi(N)]^2 + [\Psi(N)]^2\}^{\frac{1}{2}}}D_P.$$

Since this must be the same as $\dfrac{4F}{\pi}$, a condition for determining n_d is obtained in the form

$$\frac{2\pi^3 N^3 n_d M_s}{F} D_P = \{[\Phi(N)]^2 + [\Psi(N)]^2\}^{\frac{1}{2}}.$$

If D_P is measured in feet and F and M_s are expressed in tons, the condition is

$$\left[\frac{2\pi^3 N^3 M_s D_P}{Fg}\right]^2 n_d^2 = [\Phi(N)]^2 + [\Psi(N)]^2.$$

This gives a quadratic for determining n_d, the positive root being the value required.

When n_d has been determined, δ_0 can be found from the equation

$$\delta_0 = \frac{Fg\{(N^2 - n_s^2)^2 + (2Nn_d)^2\}^{\frac{1}{2}}}{2\pi^3 N^3 n_d M_s} = D_P \cdot \frac{F}{\pi P} \cdot \frac{M_G}{M_s} \cdot \frac{n_0^2\{(N^2 - n_s^2)^2 + (2Nn_d)^2\}^{\frac{1}{2}}}{N^3 n_d}.$$

For the synchronous case, when $N = N_2$ and the oscillations are most pronounced, $\Phi(N_2) = 0$, and the equation for n_d takes the simplified form

$$\frac{2\pi^3 N_2^3 M_s D_P}{Fg} n_d = \pm \Psi(N_2).$$

Taking the negative sign on the right-hand side of the equation, since it alone gives a positive value for n_d, the condition takes the form

$$\frac{\pi^3 N_2^2 M_s D_P}{Fg} n_d = \frac{N_2^2 - n_2^2}{n_2^2} n_d + \frac{N_2^2 - n_s^2}{n_0^2} n_b.$$

Hence

$$n_d = \frac{\dfrac{N_2^2 - n_s^2}{n_0^2} n_b}{\dfrac{\pi^3 N_2^2 M_s D_P}{Fg} - \dfrac{N_2^2 - n_2^2}{n_2^2}},$$

and δ_0 can then be found from the relation

$$\delta_0 = \frac{\{(N_2^2 - n_s^2)^2 + (2N_2 n_d)^2\}^{\frac{1}{2}}}{2\pi^3 N_2^3 n_d M_s} Fg.$$

A further simplification in the expression for δ_0 can be effected by taking advantage of the fact that $(2N_2 n_d)^2$ is, in all practical cases, small in comparison with $(N_2^2 - n_s^2)^2$. On this assumption an excellent approximation to δ_0 is given by

$$\delta_0 = \frac{Fg(N_2^2 - n_s^2)}{2\pi^3 N_2^3 n_d M_s},$$

and, since

$$D_P = \frac{Pg}{2\pi^2 n_0^2 M_G},$$

$$\delta_0 = \frac{F}{\pi P} \cdot \frac{n_0^2 M_G}{N_2^2 M_s} \cdot \frac{N_2^2 - n_s^2}{N_2 n_d} \cdot D_P,$$

inserting the value of n_d already determined,

$$\frac{N_2{}^2-n_s{}^2}{N_2 n_d} = \frac{n_0{}^2}{2N_2 n_b}\left[\frac{2\pi^3 N_2{}^2 M_S D_P}{Fg} - \frac{2(N_2{}^2-n_2{}^2)}{n_2{}^2}\right]$$

$$= \frac{n_0{}^2}{2N_2 n_b}\left[\frac{\pi P}{F}\cdot\frac{N_2{}^2 M_S}{n_0{}^2 M_G} - \frac{2(N_2{}^2-n_2{}^2)}{n_2{}^2}\right],$$

and, consequently,

$$\delta_0 = D_P\frac{n_0{}^2}{2N_2 n_b}\left[1 - \frac{2F}{\pi P}\cdot\frac{n_0{}^2 M_G}{n_2{}^2 M_S}\cdot\frac{N_2{}^2-n_2{}^2}{N_2{}^2}\right].$$

This again admits of simplification; since

$$n_1{}^2 = n_0{}^2\frac{M_G}{M_G+2M_U} \quad\text{and}\quad n_2{}^2 = n_0{}^2\frac{M_G}{M_G+2M_U+2M_S},$$

it follows that

$$\frac{n_0{}^2 M_G}{n_2{}^2 M_S} = \frac{2n_1{}^2}{n_1{}^2-n_2{}^2}.$$

Also, since

$$N_2{}^4 - N_2{}^2 n_1{}^2\left(1+\frac{n_s{}^2}{n_2{}^2}\right) + n_1{}^2 n_s{}^2 = 0,$$

$$N_2{}^4(n_1{}^2-n_2{}^2) = n_1{}^2[(N_2{}^2-n_s{}^2)(N_2{}^2-n_2{}^2)],$$

that is,

$$\frac{N_2{}^2-n_2{}^2}{N_2{}^2} = \frac{N_2{}^2}{N_2{}^2-n_s{}^2}\cdot\frac{n_1{}^2-n_2{}^2}{n_1{}^2}.$$

Hence, for the case when $N = N_2$, the expression for the dynamic increment of deflection at the centre of the bridge takes the comparatively simple form

$$\delta_0 = D_P\frac{n_0{}^2}{2N_2 n_b}\left[1 - \frac{4F}{\pi P}\cdot\frac{N_2{}^2}{N_2{}^2-n_s{}^2}\right].$$

For the case of two similar locomotives situated at equal distances $\frac{d}{2}$ on either side of the centre, skidding their wheels so that their hammer-blows are in unison and of equal magnitude $P\sin 2\pi Nt$, it can be shewn in a similar manner that δ_0, the maximum central deflection, is given by the expression

$$\delta_0 = 2D_P\cos\frac{\pi d}{2l}\cdot\frac{n_0{}^2}{2N_2 n_b}\left[1 - \frac{4F}{\pi P}\cdot\frac{N_2{}^2}{N_2{}^2-n_s{}^2}\right],$$

where $N_2{}^2$ is the greater of the two roots of the quadratic equation

$$N^4 - N^2 n_1{}^2\left(1+\frac{4n_b n_d}{n_0{}^2}+\frac{n_s{}^2}{n_2{}^2}\right) + n_1{}^2 n_s{}^2 = 0,$$

in which

$$n_1{}^2 = n_0{}^2\frac{M_G}{M_G+4M_U\cos^2\dfrac{\pi d}{2l}} \quad\text{and}\quad n_2{}^2 = n_0{}^2\frac{M_G}{M_G+4(M_U+M_S)\cos^2\dfrac{\pi d}{2l}}.$$

D_P is the steady central deflection due to a single force P, the hammer-blow at N_2 revs. per sec., statically applied at the centre of the bridge.

If the speed thus determined exceeds that attainable in practice, the maximum state of oscillation is developed when N has the highest permissible speed, say $N = 6$.

Following the process adopted in the case of the single central skidding locomotive, it can be shewn that for two locomotives the central dynamic deflection corresponding to a speed N and hammer-blow $P \sin 2\pi N t$ is given by

$$\delta_0 = D_P \cdot \frac{F}{\pi P \cos \frac{\pi d}{2l}} \cdot \frac{M_G}{M_s} \cdot \frac{n_0^2 \{(N^2 - n_s^2)^2 + (2Nn_d)^2\}^{\frac{1}{2}}}{N^3 n_d},$$

where n_d is the positive root of the quadratic equation

$$\left[\frac{4\pi^3 N^3 M_s D_P \cos^2 \frac{\pi d}{2l}}{Fg} \right]^2 n_d^2 = [\Phi(N)]^2 + [\Psi(N)]^2.$$

In this equation F and P are both measured in tons, M_s is measured in tons and D_P, the steady central deflection due to a force P statically applied at the centre of the bridge, is measured in feet:

$$\Phi(N) = \frac{N^4}{n_1^2} - N^2 \left(1 + \frac{4 n_b n_d}{n_0^2} + \frac{n_s^2}{n_2^2} \right) + n_s^2,$$

$$\Psi(N) = 2N \left(n_d \frac{n_2^2 - N^2}{n_2^2} + n_b \frac{n_s^2 - N^2}{n_0^2} \right),$$

$$n_1^2 = n_0^2 \frac{M_G}{M_G + 4 M_U \cos^2 \frac{\pi d}{2l}} \quad \text{and} \quad n_2^2 = n_0^2 \frac{M_G}{M_G + 4(M_U + M_s) \cos^2 \frac{\pi d}{2l}}.$$

The foregoing analysis will now be illustrated by applying it to the numerical example which was employed in the previous Chapter, and for which the data are as follows:

$$M_G = 300 \text{ tons}, \quad n_0 = 6, \quad n_b = 0\cdot36, \quad M_s = 80 \text{ tons}, \quad M_U = 20 \text{ tons},$$
$$n_s = 3, \quad P = 0\cdot6 N^2 \text{ tons}, \quad n_1^2 = 31\cdot7647, \quad n_2^2 = 21\cdot6000.$$

The deflection-frequency curves for five cases will be considered.

Case (1) Springs perfectly free.

,, (2) $F = 5$ tons

,, (3) $F = 8$ tons.

,, (4) $F = 10$ tons.

,, (5) Springs locked.

The higher critical frequency, as obtained by solving the equation

$$N^4 - N^2 n_1^2 \left(1 + \frac{n_s^2}{n_2^2}\right) + n_1^2 n_s^2,$$

gives $N_2 = 6 \cdot 11$.

This value is not absolutely correct, since N_2 depends slightly on the value of n_d, and if $n_b = 0 \cdot 36$, the correct value for N_2 is $6 \cdot 18$, but the error consequent on neglecting the influence which n_d has on the value of N_2 is of no practical importance

If the springs remain locked, the connection between the central deflection δ_0 and N, the frequency of the hammer-blow, $0 \cdot 6 N^2$ tons, is given by

$$\delta_0 = \frac{0 \cdot 6 N^2 g}{2\pi^2 n_0^2 M_G \left[\left(1 - \frac{N^2}{n_2^2}\right)^2 + \left(\frac{2Nn_b}{n_0^2}\right)^2\right]^{\frac{1}{2}}} \text{ feet,}$$

where M_G is expressed in tons.

Spring movement must necessarily take place if $4\pi^2 N^2 M_S \delta_0 > Fg$, where M_S and F are measured in tons, that is, if

$$N^4 > \frac{F}{1 \cdot 2} \cdot \frac{M_G}{M_S} n_0^2 \left[\left(1 - \frac{N^2}{n_2^2}\right)^2 + \left(\frac{2Nn_b}{n_0^2}\right)^2\right]^{\frac{1}{2}}.$$

Accordingly, if

$F = 5$ tons, spring movement occurs when N exceeds $3 \cdot 8$.

$F = 8$ tons,　　　 ,,　　　　 ,,　　　 N　 ,,　　 $3 \cdot 9$.

$F = 10$ tons,　　 ,,　　　　 ,,　　　 N　 ,,　　 $4 \cdot 1$.

Case (1) *Spring movement free from friction.*

The deflection-frequency curve for this case is obtained from the general formula established in Chapter IX by putting $n_d = 0$.

This gives
$$\delta_0 = D_P \frac{N^2 - n_s^2}{\{[\Phi(N)]^2 + [\Psi(N)]^2\}^{\frac{1}{2}}} = K D_P,$$

where
$$\Phi(N) = \frac{N^4}{n_1^2} - N^2 \left(1 + \frac{n_s^2}{n_2^2}\right) + n_s^2$$

and
$$\Psi(N) = 2 N n_b \frac{n_s^2 - N^2}{n_0^2}.$$

For values of N, ranging from 0 to 10, the values of D_P, $\Phi(N)$, $\Psi(N)$, $N^2 - n_s^2$, K and δ_0 are tabulated below and the deflection-frequency curve is given by Fig. 34.

It will be seen that peak values occur when $N = 6 \cdot 11$ and $2 \cdot 77$, these being the two positive roots of the equation $\Phi(N) = 0$.

As N increases from 2·77 to 3, the deflection abruptly falls to zero, and this is a merciful circumstance for the springs, since the slightest prolonged

N	D_P ins.	$\Phi(N)$	$\Psi(N)$	$N^2 - n_s^2$	K	δ_0 ins.
0	0·0000	9·0000	0·0000	− 9·0000	1·0000	0·0000
1	0·0011	7·6148	0·1600	− 8·0000	1·0504	0·0012
2	0·0044	3·8370	0·2000	− 5·0000	1·3013	0·0057
2·5	0·0068	1·3756	0·1375	− 2·7500	1·9890	0·0135
2·7	0·0079	0·3456	0·0923	− 1·7100	4·7815	0·0378
2·77	0·0083	0·0000	0·0745	− 1·3448	18·0510	0·1498
2·8	0·0085	− 0·1716	0·0650	− 1·1600	6·4846	0·0551
3·0	0·0098	− 1·2000	0·0000	0·0000	0·0000	0·0000
3·5	0·0133	− 3·6300	− 0·2275	3·2500	0·8936	0·0119
4	0·0174	− 5·6074	− 0·5600	7·0000	1·2422	0·0216
4·5	0·0220	− 6·7781	− 1·0125	11·2500	1·3864	0·0305
5	0·0272	− 6·7407	− 1·6000	16·0000	2·3339	0·0635
5·5	0·0329	− 5·0466	− 2·3375	21·2500	3·8208	0·1257
6	0·0391	− 1·2000	− 3·2400	27·0000	7·8146	0·3056
6·11	0·0406	0·0000	− 3·4622	28·3321	8·1833	0·3322
6·5	0·0459	5·3423	− 4·3225	33·2500	4·8385	0·2221
7	0·0533	15·1704	− 5·6000	40·0000	2·4735	0·1299
8	0·0696	47·2816	− 8·0000	55·0000	1·1436	0·0796
9	0·0880	100·8001	− 12·9600	72·0000	0·7085	0·0623
10	0·1087	173·1481	− 18·2000	91·0000	0·5227	0·0568

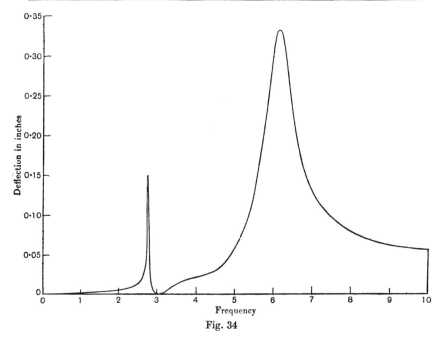

Fig. 34

oscillation of the bridge at a frequency of 3 would induce movements in the spring-borne mass of a resonant character and the deflection in the springs would mount up to an indefinite extent.

Even when $N = 2\cdot77$, the approach to this condition of synchronism is sufficiently close to induce large spring movements, and it is very evident that, without adequate damping, the riding of a locomotive at some speeds would be dangerous, by reason of the violent oscillations developed in its springs.

Cases (2), (3) and (4), in which F = 5, 8 and 10 tons respectively.

To deduce the deflection-frequency curves for these cases, the first step is to calculate the values of n_d corresponding to the three values of F for the range of values of N, which has been taken from $N = 0$ to $N = 10$. The condition for determining n_d is the quadratic equation previously established, viz.

$$\left[\frac{2\pi^3 N^3 M_S D_P}{Fg}\right]^2 n_d{}^2 = [\Phi(N)]^2 + [\Psi(N)]^2,$$

where F and M_S are measured in tons and D_P, the central deflection due to a steady central force $P = 0.6N^2$ tons, is given by

$$D_P = \frac{0.6N^2 g}{2\pi^2 n_0{}^2 M_G} \text{ feet.}$$

Having obtained n_d, the central dynamical deflection is obtained from the formula

$$\delta_0 = \frac{[(N^2 - n_s{}^2)^2 + (2Nn_d)^2]^{\frac{1}{2}}}{2\pi^3 N^3 n_d M_S} Fg.$$

The results obtained by this process are set forth in the following table, and Fig. 35 gives the deflection-frequency curves for these three cases.

N	P in tons	D_P in ins.	n_d	Dynamic magnifier	Deflection in ins.
				Case (2) $F = 5$ tons	
4·5	12·15	0·0220	1·562	2·199	0·048
5·0	15·00	0·0272	0·949	2·244	0·061
5·5	18·15	0·0329	0·382	4·075	0·134
6·0	21·60	0·0391	0·253	4·946	0·193
6·11	22·407	0·0406	0·233	5·129	0·208
6·5	25·35	0·0459	0·286	3·617	0·166
7·0	29·40	0·0533	0·411	2·093	0·122
8·0	38·40	0·0696	0·581	1·049	0·073
9·0	48·60	0·0880	0·659	0·671	0·059
10·0	60·00	0·1087	0·689	0·479	0·052

n_d	Dynamic magnifier	Deflection in ins.	n_d	Dynamic magnifier	Deflection in ins.	N
Case (3) $F = 8$ tons			Case (4) $F = 10$ tons			
2·321	3·149	0·069	3·878	3·672	0·081	4·5
1·747	2·484	0·068	2·445	2·735	0·074	5·0
0·810	2·894	0·095	1·368	2·707	0·089	5·5
0·630	3·282	0·128	1·255	2·317	0·091	6·0
0·588	3·345	0·136	1·194	2·243	0·091	6·11
0·644	2·629	0·121	1·216	1·863	0·086	6·5
0·834	1·701	0·091	1·402	1·353	0·072	7·0
1·077	0·935	0·065	1·596	0·830	0·058	8·0
1·165	0·625	0·055	1·619	0·582	0·051	9·0
1·177	0·457	0·050	1·578	0·437	0·047	10·0

Case (5) *Springs held locked by friction.*

The central dynamical deflection in this case is given by

$$\delta_0 = \frac{0 \cdot 6 N^2 g}{2\pi^2 n_0^2 M_G \left[\left(1 - \dfrac{N^2}{n_2^2}\right)^2 + \left(\dfrac{2 N n_b}{n_0^2}\right)^2\right]^{\frac{1}{2}}} \text{ feet.}$$

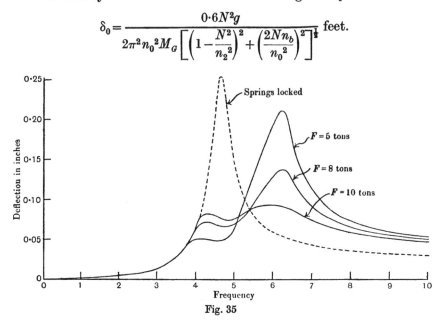

Fig. 35

The crest value occurs when $N = n_2 = 4 \cdot 63$, and the values of δ_0, measured in inches, for values of N ranging from $N = 0$ to $N = 10$ are tabulated below:

N	0	1	2	3	4	4·63	5	6	7	8	9	10
δ_0	0	0·001	0·005	0·017	0·064	0·252	0·145	0·058	0·042	0·035	0·032	0·030

The deflection-frequency curve for this case is shewn (dotted) in Fig. 35.

From an examination of the curves for the cases $F = 5$, 8 and 10 shewn in Fig. 35, it appears approximately that their crests all occur when N has the value 6·11 approximately, as derived from the equation

$$N^4 - N^2 n_1^2 \left(1 + \frac{n_s^2}{n_2^2}\right) + n_1^2 n_s^2 = 0,$$

but that their heights vary greatly, decreasing in value as F increases. A further increase in F will, however, cause the deflection to rise again consequent on the point of departure from the dotted curve mounting up towards the apex at $N = 4 \cdot 63$, and, for this particular bridge and locomotive, $F = 10$ is approximately the most advantageous value for the spring damping.

Fig. 35 indicates very clearly the important part F plays in prescribing bridge oscillations, and the necessity for obtaining information, more precise

than at present exists, concerning this somewhat elusive locomotive characteristic. One method for doing this is to measure the vertical movements
of the locomotive on its springs for increasing and decreasing loads, and the
intercept between the loading and unloading curves will then give a measure
of the friction in the spring movement. Alternatively the locomotive could
be placed on an oscillating platform such as a bridge set in motion by an
independent oscillator. By observing the deflections thus induced in the
springs the friction in the spring mechanism could be deduced. No doubt
some portion of this friction is due to the pressure between the axle-boxes
and their guides, and, as such, is affected by the thrust in the connecting-
rods and side-rods due to steam pressure and inertia. Accordingly, F for a
given locomotive can hardly be treated as having a constant value, but
rather a value ranging between an upper and lower limit which can only be
determined by experiments designed for the purpose.

Using the approximate formula

$$\delta_0 = D_P \frac{n_0{}^2}{2N_2 n_b} \left[1 - \frac{4F}{\pi P} \frac{N_2{}^2}{N_2{}^2 - n_s{}^2} \right],$$

which gives the dynamic magnification for the extreme case when $N = N_2$,
the results for Cases (2), (3) and (4) are 5·12, 3·28 and 2·06 respectively as
against 5·13, 3·35 and 2·24, as obtained by the more detailed analysis,
shewing that this comparatively simple method of obtaining a result of
great practical importance leaves little to be desired in the matter of
accuracy.

CHAPTER XI

VECTOR METHODS FOR COMPUTING OSCILLATIONS DUE TO ALTERNATING FORCES

The graphical methods developed in this Chapter are based upon the vector method for determining the particular integral of a linear differential equation whose right-hand side is a periodic function of time, and the method will be illustrated by finding the particular integral of the equation

$$a\frac{d^2x}{dt^2} + b\frac{dx}{dt} + cx = P\cos 2\pi Nt \qquad \ldots\ldots(1),$$

where a, b and c are constants.

The integral will be of the general form $x = x_0 \cos(2\pi Nt - \alpha)$, and the problem resolves itself into the determination of the values of x_0 and α.

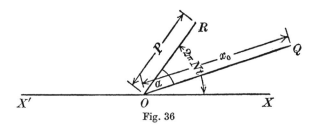

Fig. 36

$P\cos 2\pi Nt$, the right-hand side of equation (1), can be viewed (see Fig. 36) as the projection, on a base XOX', of a vector OR of length P, rotating about O with uniform angular velocity $2\pi N$, OX being the direction of OP when $t = 0$.

x, in the same way, can be viewed as the projection of a vector OQ of length x_0, lagging an angle α behind OR.

$\dfrac{dx}{dt} = 2\pi N x_0 \cos\left(2\pi Nt + \dfrac{\pi}{2} - \alpha\right)$, and, as such, can be represented by the projection of a vector of length $2\pi N x_0$, 90° in advance of the vector for x.

$\dfrac{d^2x}{dt^2} = 4\pi^2 N^2 x_0 \cos(2\pi Nt + \pi - \alpha)$ can be represented by the projection of a vector of length $4\pi^2 N^2 x_0$, 180° in advance of the vector for x.

Hence the three terms constituting the left-hand side of equation (1) can be viewed as the projections of vectors OA, AB and BC, where $OA = cx_0$,

$AB = 2\pi Nbx_0$, 90° in advance of OA, and $BC = 4\pi^2N^2ax_0$, 180° in advance of OA (see Fig. 37).

The sum of the three terms on the left-hand side of equation (1) is accordingly given by the projection of the vector OC, whose length is

$$x_0\{(c - 4\pi^2N^2a)^2 + (2\pi Nb)^2\}^{\frac{1}{2}},$$

and whose direction α with reference to OA is defined by

$$\tan\alpha = \frac{2\pi Nb}{c - 4\pi^2N^2a}.$$

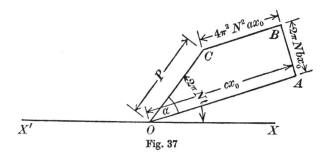

Fig. 37

Equation (1) accordingly states that OC is identical with the vector for $P\cos 2\pi Nt$, in other words, $OC = P$ and angle XOC is $2\pi Nt$.

Hence
$$x_0 = \frac{P}{\{(c - 4\pi^2N^2a)^2 + (2\pi Nb)^2\}^{\frac{1}{2}}}$$

and
$$x = \frac{P\cos(2\pi Nt - \alpha)}{\{(c - 4\pi^2N^2a)^2 + (2\pi Nb)^2\}^{\frac{1}{2}}},$$

where
$$\tan\alpha = \frac{2\pi Nb}{c - 4\pi^2N^2a}.$$

This method can be extended to any linear differential equation with constant coefficients for which the right-hand side is a periodic function of t, and it is the easiest method for writing down the particular integrals for the numerous equations of this type which have appeared in the previous Chapters.

This vector method will now be employed to determine the state of oscillation set up in a bridge by a stationary locomotive which is skidding its wheels at a constant speed.

The force responsible for the oscillations is the reaction between the rails and the driving-wheels of the locomotive. If the speed of rotation of the wheels is N per second, then after the motion has become steady this reaction can be written $R \sin 2\pi N t$, and this will in future be termed the " Operative Force".

If the locomotive is stationed at the centre of the bridge the fundamental harmonic component of the operative force is $\frac{2}{l} R \sin 2\pi N t \sin \frac{\pi x}{l}$, and the equation for the oscillation thus produced is

$$EI \frac{d^4 y}{dx^4} + 4\pi n_b m \frac{dy}{dt} + m \frac{d^2 y}{dt^2} = \frac{2}{l} R \sin 2\pi N t \sin \frac{\pi x}{l}.$$

The particular integral of this is of the form $y = f(t) \sin \frac{\pi x}{l}$, where

$$\frac{d^2 f}{dt^2} + 4\pi n_b \frac{df}{dt} + 4\pi^2 n_0^2 f(t) = \frac{2}{M_G} R \sin 2\pi N t,$$

and consequently

$$f(t) = \frac{\lfloor R \sin (2\pi N t - \alpha)}{2\pi^2 n_0^2 M_G \left\{ \left(1 - \frac{N^2}{n_0^2}\right)^2 + \left(\frac{2N n_b}{n_0^2}\right)^2 \right\}^{\frac{1}{2}}},$$

where

$$\tan \alpha = \frac{2N n_b}{n_0^2 - N^2}.$$

This result is equivalent to saying that, if δ, the central deflection, is written $\delta = \delta_0 \sin (2\pi N t - \alpha)$, then δ is given by the projection of a vector of length

$$\frac{R}{2\pi^2 n_0^2 M_G \left\{ \left(1 - \frac{N^2}{n_0^2}\right)^2 + \left(\frac{2N n_b}{n_0^2}\right)^2 \right\}^{\frac{1}{2}}},$$

which lags an angle $\tan^{-1} \frac{2N n_b}{n_0^2 - N^2}$ behind the vector for the Operative Force.

Conversely, the operative force is given by the projection of a vector OP of length

$$2\pi^2 n_0^2 M_G \left\{ \left(1 - \frac{N^2}{n_0^2}\right)^2 + \left(\frac{2N n_b}{n_0^2}\right)^2 \right\}^{\frac{1}{2}} \delta_0,$$

moving in advance of OQ, the vector for the deflection, by an angle

$$\tan^{-1} \frac{2N n_b}{n_0^2 - N^2},$$

as indicated in Fig. 38.

If there are two locomotives situated at distances a_1 and a_2 from the end of the span, the two operating forces being

$$R_1 \sin (2\pi N t + \phi_1) \quad \text{and} \quad R_2 \sin (2\pi N t + \phi_2),$$

the equivalent central Operative Force is

$$R_1 \sin (2\pi N t + \phi_1) \sin \frac{\pi a_1}{l} + R_2 \sin (2\pi N t + \phi_2) \sin \frac{\pi a_2}{l}.$$

The vector for this must have the value

$$2\pi^2 n_0{}^2 M_G \left\{ \left(1 - \frac{N^2}{n_0{}^2}\right)^2 + \left(\frac{2Nn_b}{n_0{}^2}\right)^2 \right\}^{\frac{1}{2}} \delta_0,$$

and must lead the vector for the central deflection by an angle

$$\tan^{-1} \frac{2Nn_b}{n_0{}^2 - N^2}.$$

The operative force, R_1, for a single locomotive will now be analysed and viewed as the projection of three different vectors, one depending on the

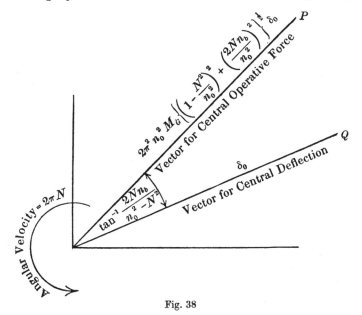

Fig. 38

hammer-blow P, another depending on F, the frictional forces resisting spring movement, and the third depending on the magnitude of the central deflection.

The notation employed will be that specified in Chapter IX.

For a locomotive situated at a distance a_1 from the end of the bridge, at time t, let $\quad z_1$ be the depth of M_S below its mean level

$\qquad h_1 \quad ,, \quad ,, \quad M_U \quad ,, \quad ,,$

The extra thrust in the springs, due to their compression $z_1 - h_1$, can be written $4\pi^2 n_s{}^2 M_S (z_1 - h_1)$, and, on the assumption discussed in Chapter X that the frictional force can be replaced by the first harmonic component, the downward thrust on M_U due to friction is $\dfrac{4F}{\pi} \sin 2\pi N (t - t_1)$, where t_1, the phase difference between the motion of the bridge and the spring movement, has to be found from the condition that the friction is momentarily zero when there is no spring movement, that is, $\dfrac{d}{dt}(z_1 - h_1) = 0$, when $t = t_1$. By considering the motion of M_S,

$$M_S \frac{d^2}{dt^2}(z_1 - h_1) + 4\pi^2 n_s{}^2 M_S (z_1 - h_1) = - M_S \frac{d^2 h_1}{dt^2} - \frac{4F}{\pi} \sin 2\pi N (t - t_1)$$

and, since $h_1 = \delta_0 \sin \dfrac{\pi a_1}{t} \sin 2\pi Nt$, this becomes

$$\frac{d^2}{dt^2}(z_1 - h_1) + 4\pi^2 n_s{}^2 (z_1 - h_1) = 4\pi^2 N^2 \delta_0 \sin \frac{\pi a_1}{l} \sin 2\pi Nt - \frac{4F}{\pi M_S} \sin 2\pi N (t - t_1).$$

Hence

$$z_1 - h_1 = - \frac{N^2}{N^2 - n_s{}^2} \delta_0 \sin \frac{\pi a_1}{l} \sin 2\pi Nt + \frac{4F}{\pi M_S} \cdot \frac{\sin 2\pi N (t - t_1)}{4\pi^2 (N^2 - n_s{}^2)},$$

and the condition $\dfrac{d}{dt}(z_1 - h_1) = 0$, when $t = t_1$, gives

$$\cos 2\pi Nt_1 = \frac{\dfrac{4F}{\pi}}{4\pi^2 N^2 M_S \delta_0 \sin \dfrac{\pi a_1}{l}}.$$

If the hammer-blow for the locomotive is $P \sin (2\pi Nt + \gamma)$, the corresponding operative force has the composition given by

$$R_1 = P \sin (2\pi Nt + \gamma) + \frac{4F}{\pi} \sin 2\pi N (t - t_1) + 4\pi^2 n_s{}^2 M_S (z_1 - h_1).$$

Substituting the value of $z_1 - h_1$ just determined, the following result is obtained:

$$R_1 = P \sin (2\pi Nt + \gamma) + \frac{4F}{\pi} \cdot \frac{N^2}{N^2 - n_s{}^2} \sin 2\pi N (t - t_1)$$
$$- \frac{4\pi^2 N^2}{N^2 - n_s{}^2} [(M_U + M_S) n_s{}^2 - M_U N^2] \delta_0 \sin \frac{\pi a_1}{l} \sin 2\pi Nt.$$

For a similar skidding locomotive with its hammer-blow in step with the first locomotive, but situated at a distance a_2 from the end of the bridge, the operative force is given by

$$R_2 = P \sin (2\pi Nt + \gamma) + \frac{4F}{\pi} \cdot \frac{N^2}{N^2 - n_s{}^2} \sin 2\pi N (t - t_2)$$
$$- \frac{4\pi^2 N^2}{N^2 - n_s{}^2} [(M_U + M_S) n_s{}^2 - M_U N^2] \delta_0 \sin \frac{\pi a_2}{l} \sin 2\pi Nt,$$

where
$$\cos 2\pi Nt_2 = \frac{\dfrac{4F}{\pi}}{4\pi^2 N^2 M_S \delta_0 \sin \dfrac{\pi a_2}{l}}.$$

The equivalent central operative force, $\left[R_1 \sin \frac{\pi a_1}{l} + R_2 \sin \frac{\pi a_2}{l} \right]$, accordingly takes the form

$$
P \sin (2\pi N t + \gamma) \left[\sin \frac{\pi a_1}{l} + \sin \frac{\pi a_2}{l} \right]
$$

$$
+ \frac{4F}{\pi} \cdot \frac{N^2}{N^2 - n_s^2} \left[\sin 2\pi N (t - t_1) \sin \frac{\pi a_1}{l} + \sin 2\pi N (t - t_2) \sin \frac{\pi a_2}{l} \right]
$$

$$
- \frac{4\pi^2 N^2}{N^2 - n_s^2} [(M_U + M_S) n_s^2 - M_U N^2] \left[\sin^2 \frac{\pi a_1}{l} + \sin^2 \frac{\pi a_2}{l} \right] \delta_0 \sin 2\pi N t.
$$

It has been shewn that the vector for this central operative force has the value

$$
2\pi^2 n_0^2 M_G \left\{ \left(1 - \frac{N^2}{n_0^2} \right)^2 + \left(\frac{2Nn_b}{n_0^2} \right)^2 \right\}^{\frac{1}{2}} \delta_0,
$$

and leads the deflection vector by an angle $\tan^{-1} \dfrac{2Nn_b}{n_0^2 - N^2}$, and, from this vector equivalence, a figure can be constructed from which the numerical value of δ_0, for any particular case, can be scaled off.

VECTOR DIAGRAM FOR DETERMINING DEFLECTION

Let OQ be the vector for the central deflection $\delta_0 \sin 2\pi N t$. (See Fig. 39.)
Let OA be the vector for

$$
- \frac{4\pi^2 N^2}{N^2 - n_s^2} [(M_U + M_S) n_s^2 - M_U N^2] \left[\sin^2 \frac{\pi a_1}{l} + \sin^2 \frac{\pi a_2}{l} \right] \delta_0 \sin 2\pi N t.
$$

Let OD be the vector for the central operative force, that is, set off

$$
OD = 2\pi^2 n_0^2 M_G \left\{ \left(1 - \frac{N^2}{n_0^2} \right)^2 + \left(\frac{2Nn_b}{n_0^2} \right)^2 \right\}^{\frac{1}{2}} \delta_0
$$

at an angle $\tan^{-1} \dfrac{2Nn_b}{n_0^2 - N^2}$ in advance of OQ.

On a base AB_1, of length

$$
\frac{2N^2}{N^2 - n_s^2} \cdot 4\pi^2 N^2 M_S \sin^2 \frac{\pi a_1}{l} \delta_0,
$$

describe a semi-circle.

On a base AB_2, of length

$$
\frac{2N^2}{N^2 - n_s^2} \cdot 4\pi^2 N^2 M_S \sin^2 \frac{\pi a_2}{l} \delta_0,
$$

describe a semi-circle.

Set off
$$AC_1 = \frac{8F}{\pi} \cdot \frac{N^2}{N^2 - n_s^2} \sin \frac{\pi a_1}{l},$$

C_1 lying on the semi-circle whose base is AB_1.

Set off
$$AC_2 = \frac{8F}{\pi} \cdot \frac{N^2}{N^2 - n_s^2} \sin \frac{\pi a_2}{l},$$

C_2 lying on the semi-circle whose base is AB_2.

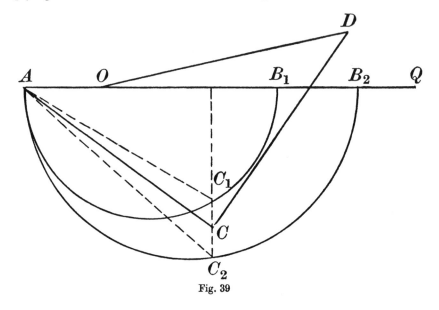

Fig. 39

Then
$$\cos C_1 A B_1 = \frac{\dfrac{4F}{\pi}}{4\pi^2 N^2 M_s \delta_0 \sin \dfrac{\pi a_1}{l}} = \cos 2\pi N t_1,$$

also
$$\cos C_2 A B_2 = \frac{\dfrac{4F}{\pi}}{4\pi^2 N^2 M_s \delta_0 \sin \dfrac{\pi a_2}{l}} = \cos 2\pi N t_2,$$

and, since $AC_1 \cos C_1 A B_1 = AC_2 \cos C_2 A B_2$, it follows that C_1 and C_2 are in a vertical line.

Let C be the middle point of $C_1 C_2$.

Since $2\pi N t_1 = C_1 A B_1$ and $2\pi N t_2 = C_2 A B_2$, it follows that $\frac{1}{2}AC_1$ is the vector for

$$\frac{4F}{\pi} \cdot \frac{N^2}{N^2 - n_s^2} \sin \frac{\pi a_1}{l} \sin 2\pi N (t - t_1),$$

and $\frac{1}{2}AC_2$ is the vector for

$$\frac{4F}{\pi} \cdot \frac{N^2}{N^2 - n_s^2} \sin \frac{\pi a_2}{l} \sin 2\pi N (t - t_2),$$

and, consequently, AC is the vector for

$$\frac{4F}{\pi} \cdot \frac{N^2}{N^2 - n_s^2} \left[\sin \frac{\pi a_1}{l} \sin 2\pi N (t - t_1) + \sin \frac{\pi a_2}{l} \sin 2\pi N (t - t_2) \right].$$

Hence the vector equivalence stated above shews that the closing line CD must be the vector for

$$P \sin (2\pi N t + \gamma) \left[\sin \frac{\pi a_1}{l} + \sin \frac{\pi a_2}{l} \right].$$

Although δ_0 is not known to start with, the relative magnitudes and directions of OA, OD, AB_1 and AB_2 can be determined. The magnitude of the hammer-blow P and the value of the friction force F being given, the ratio of AC_1 or AC_2 to CD is known, and this condition enables the position of C to be located. The magnitude of CD is $P \left(\sin \frac{\pi a_1}{l} + \sin \frac{\pi a_2}{l} \right)$, and this being known, the scale of the diagram and consequently the value of δ_0 can thus be determined.

This vector method will now be illustrated by the following numerical example:

Bridge characteristics.

Span 120 feet, $M_G = 200$ tons, $n_0 = 7$, $n_b = 0.4$.

Locomotive characteristics.

$M_U = 20$ tons, $M_S = 60$ tons, $n_s = 3$, $N = 5$, $F = 10$ tons, $P = 15$ tons at 5 revs. per sec.

The bridge is oscillated by two similar skidding locomotives with their hammer-blows in phase, and stationed at distances of 30 feet and 80 feet from one end of the bridge.

Accordingly,

$$\sin \frac{\pi a_1}{l} = \sin \frac{\pi}{4} = 0.70711, \quad \sin^2 \frac{\pi a_1}{l} = 0.5,$$

$$\sin \frac{\pi a_2}{l} = \sin \frac{2\pi}{3} = 0.86602, \quad \sin^2 \frac{\pi a_2}{l} = 0.75.$$

The angle by which OD leads OQ is

$$\tan^{-1} \frac{2 N n_b}{n_0^2 - N^2} = \tan^{-1} \frac{1}{6}.$$

If δ_0, the central deflection, is measured in feet, and the masses are measured in tons:

OD scales $2\pi^2 M_G \{(n_0{}^2 - N^2)^2 + (2Nn_b)^2\}^{\frac{1}{2}} \dfrac{\delta_0}{g}$ tons $= 2985\delta_0$ tons,

OA scales $\dfrac{4\pi^2 N^2}{N^2 - n_s{}^2}[(M_U + M_S)n_s{}^2 - M_U N^2]$
$$\times \left[\sin^2\frac{\pi a_1}{l} + \sin^2\frac{\pi a_2}{l}\right]\frac{\delta_0}{g} \text{ tons} = 527\delta_0 \text{ tons,}$$

AB_1 scales $\dfrac{2N^2}{N^2 - n_s{}^2} \cdot 4\pi^2 N^2 M_S \sin^2\dfrac{\pi a_1}{l} \cdot \dfrac{\delta_0}{g}$ tons $= 2875\delta_0$ tons,

AB_2 scales $\dfrac{2N^2}{N^2 - n_s{}^2} \cdot 4\pi^2 N^2 M_S \sin^2\dfrac{\pi a_2}{l} \cdot \dfrac{\delta_0}{g}$ tons $= 4313\delta_0$ tons,

AC_1 scales $\dfrac{8F}{\pi} \cdot \dfrac{N^2}{N^2 - n_s{}^2} \sin\dfrac{\pi a_1}{l}$ tons $= 28 \cdot 135$ tons,

AC_2 scales $\dfrac{8F}{\pi} \cdot \dfrac{N^2}{N^2 - n_s{}^2} \cdot \sin\dfrac{\pi a_2}{l}$ tons $= 34 \cdot 458$ tons,

CD scales $P\left[\sin\dfrac{\pi a_1}{l} + \sin\dfrac{\pi a_2}{l}\right]$ tons $= 23 \cdot 597$ tons,

$$\frac{AC_2}{CD} = \frac{34 \cdot 458}{23 \cdot 597} = 1 \cdot 46.$$

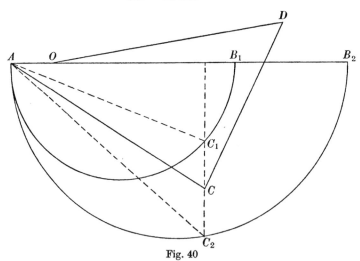

Fig. 40

Fig. 40 is the vector diagram for this case arranged so that $\dfrac{AC_2}{CD} = 1 \cdot 46$.

CD represents $23 \cdot 597$ and AB_2 to this same scale represents $44 \cdot 834$ tons. But AB_2 was set off to represent $4313\delta_0$ tons.

Hence $\delta_0 = \dfrac{44 \cdot 834}{4313}$ feet $= 0 \cdot 0104$ foot $= 0 \cdot 125$ inch.

For the case of a single locomotive at the centre of the span, the vector diagram takes the simplified form shewn in Fig. 41.

In this diagram

OD scales $2\pi^2 M_G \{(n_0{}^2 - N^2)^2 + (2Nn_b)^2\}^{\frac{1}{2}} \dfrac{\delta_0}{g}$ tons,

OA scales $\dfrac{4\pi^2 N^2}{N^2 - n_s{}^2}[(M_U + M_S)\,n_s{}^2 - M_U N^2]\dfrac{\delta_0}{g}$ tons,

AB scales $\dfrac{N^2}{N^2 - n_s{}^2} \cdot 4\pi^2 N^2 M_S \dfrac{\delta_0}{g}$ tons,

AC scales $\dfrac{4F}{\pi} \cdot \dfrac{N^2}{N^2 - n_s{}^2}$ tons,

CD scales P tons   and $\tan BOD = \dfrac{2Nn_b}{n_0{}^2 - N^2}$,

where δ_0 is measured in feet and the masses are measured in tons. For the particular case under consideration $OD = 2985\delta_0$ tons, $OA = 422\delta_0$ tons, $AB = 2875\delta_0$ tons, $AC = 19\cdot894$ tons, $CD = 15$ tons, and $\tan BOD = \frac{1}{6}$.

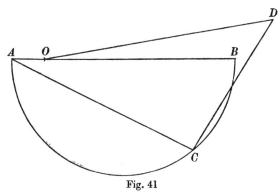

Fig. 41

$\dfrac{AC}{CD} = 1\cdot33$, and the vector diagram, Fig. 41, has been drawn to satisfy these conditions. To the scale which makes CD represent 15 tons, AB represents 22·345 tons.

Hence $2875\delta_0 = 22\cdot345$; that is, $\delta_0 = 0\cdot0078$ foot $= 0\cdot094$ inch.

For this particular case the two critical frequencies are given by the equation

$$N^4 - N^2 n_1{}^2 \left(1 + \dfrac{n_s{}^2}{n_2{}^2}\right) + n_1{}^2 n_s{}^2 = 0,$$

where

$$n_1{}^2 = n_0{}^2 \dfrac{M_G}{M_G + 2M_U} = \dfrac{245}{6}$$

and

$$n_2{}^2 = n_0{}^2 \dfrac{M_G}{M_G + 2M_U + 2M_S} = \dfrac{245}{9}.$$

From this it will be found that $N = 6\cdot81$ and $2\cdot81$ and the greatest state of oscillation will be developed when $N = 6\cdot81$.

VECTOR DIAGRAM FOR THE SYNCHRONOUS CASE

$$N = N_2 = 6\cdot81$$

OD scales $\quad 2\pi^2 M_G \{(n_0{}^2 - N_2{}^2)^2 + (2N_2 n_b)^2\}^{\frac{1}{2}} \dfrac{\delta_0}{g}$ tons $= 740 \delta_0$ tons,

OA scales $\quad \dfrac{4\pi^2 N_2{}^2}{N_2{}^2 - n_s{}^2} [(M_U + M_S) n_s{}^2 - M_U N_2{}^2] \dfrac{\delta_0}{g}$ tons $= -317 \delta_0$ tons,

AB scales $\quad \dfrac{N_2{}^2}{N_2{}^2 - n_s{}^2} \cdot 4\pi^2 N_2{}^2 M_S \dfrac{\delta_0}{g}$ tons $= 4238 \delta_0$ tons,

AC scales $\quad \dfrac{4F}{\pi} \cdot \dfrac{N_2{}^2}{N_2{}^2 - n_s{}^2}$ tons $= 15\cdot795$ tons,

CD scales $\quad 27\cdot849$ tons, $\quad \dfrac{AC}{CD} = 0\cdot567,$

and the angle by which OD leads OB is

$$\tan^{-1} \frac{2N_2 n_b}{n_0{}^2 - N_2{}^2} = \tan^{-1} 2\cdot08.$$

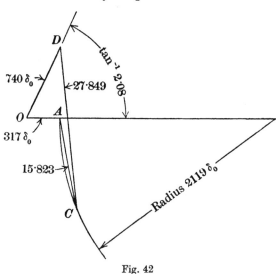

Fig. 42

The vector diagram which satisfies these conditions is shewn by Fig. 42. To the scale which makes CD represent $27\cdot849$ tons, OD will represent $13\cdot382$ tons. Since OD also represents $740 \delta_0$ tons, it follows that

$$\delta_0 = \frac{13\cdot382}{740} \text{ foot} = 0\cdot217 \text{ inch.}$$

The approximate formula established in Chapter X for this deflection, namely

$$\delta_0 = D_P \frac{n_0{}^2}{2N_2 n_b} \left[1 - \frac{4F}{\pi P} \cdot \frac{N_2{}^2}{N_2{}^2 - n_s{}^2} \right],$$

gives $\delta_0 = 0 \cdot 216$ inch, and the discrepancy is within the limits of practical draughtsmanship.

The accuracy of this approximate formula depends on the fact that when $N = N_2$, D is vertically above A, and C is very nearly in this same vertical line. The condition that D is vertically above A is

$$2\pi^2 M_G (n_0{}^2 - N_2{}^2) = \frac{4\pi^2 N_2{}^2}{N_2{}^2 - n_s{}^2} [(M_U + M_S) n_s{}^2 - M_U N_2{}^2],$$

that is, $(n_0{}^2 - N_2{}^2)(N_2{}^2 - n_s{}^2) = N_2{}^2 \left[\frac{n_0{}^2 - n_2{}^2}{n_2{}^2} n_s{}^2 - \frac{n_0{}^2 - n_1{}^2}{n_1{}^2} N_2{}^2 \right],$

and this reduces to $N_2{}^4 - N_2{}^2 n_1{}^2 \left(1 + \frac{n_s{}^2}{n_2{}^2} \right) + n_s{}^2 n_1{}^2 = 0,$

which is the equation by which the critical frequency N_2 is determined.

Hence D is vertically above A and the deviation of AC from the vertical is so slight that CD is very nearly $AD + AC$.

To this degree of accuracy

$$P = 2\pi^2 M_G . 2N_2 n_b \frac{\delta_0}{g} + \frac{4F}{\pi} . \frac{N_2{}^2}{N_2{}^2 - n_s{}^2}.$$

Hence $\delta_0 = \frac{Pg}{2\pi^2 n_0{}^2 M_G} . \frac{n_0{}^2}{2N_2 n_b} \left[1 - \frac{4F}{\pi P} . \frac{N_2{}^2}{N_2{}^2 - n_s{}^2} \right],$

that is, $\delta_0 = D_P . \frac{n_0{}^2}{2N_2 n_b} \left[1 - \frac{4F}{\pi P} . \frac{N_2{}^2}{N_2{}^2 - n_s{}^2} \right],$

and this confirms the result obtained in Chapter X, by analytical methods.

Reverting to the general case represented by Fig. 39, a position of C_1 on its semi-circle can only be found if AC_1 is less than AB_1, that is, if

$$4\pi^2 N^2 M_S \frac{\delta_0}{g} \sin \frac{\pi a_1}{l}$$

is greater than $\dfrac{4F}{\pi}$, where F and M_S are measured in tons and δ_0 in feet.

This is the condition that the inertia of the spring-borne mass is sufficient to break down friction, when the frictional force is represented by its first harmonic component, having a crest value $\dfrac{4F}{\pi}$.

Hence, if the conditions of the vector diagram can be satisfied with C_1 located on its semi-circle, it is evidence that the locomotive in question is moving on its springs. It may happen that when two locomotives are in action, spring movement is induced for one and not for the other. By a small extension of the vector diagram, such cases can be studied, and, in the process of constructing the diagram, the question of whether the springs are locked or free is settled automatically without preliminary investigation.

VECTOR DIAGRAM FOR THE CASE WHEN THE SPRINGS ARE LOCKED FOR LOCOMOTIVE No. 1 AND FREE FOR LOCOMOTIVE No. 2

The operative forces for this case are given by

$$R_1 = P\sin(2\pi Nt+\gamma) + 4\pi^2 N^2 [M_U + M_s]\frac{\delta_0}{g}\sin\frac{\pi a_1}{l}\sin 2\pi Nt$$

and

$$R_2 = P\sin(2\pi Nt+\gamma) + \frac{4F}{\pi}\cdot\frac{N^2}{N^2-n_s^2}\sin 2\pi N(t-t_1)$$

$$-\frac{4\pi^2 N^2}{N^2-n_s^2}[(M_U+M_s)n_s^2 - M_U N^2]\frac{\delta_0}{g}\sin\frac{\pi a_2}{l}\sin 2\pi Nt.$$

The equivalent central operative force $\left(R_1\sin\frac{\pi a_1}{l}+R_2\sin\frac{\pi a_2}{l}\right)$ can accordingly be expressed in the form

$$P\sin(2\pi Nt+\gamma)\left[\sin\frac{\pi a_1}{l}+\sin\frac{\pi a_2}{l}\right] - \frac{4\pi^2 N^2}{N^2-n_s^2}[(M_U+M_s)n_s^2 - M_U N^2]$$

$$\times\left[\sin^2\frac{\pi a_1}{l}+\sin^2\frac{\pi a_2}{l}\right]\frac{\delta_0}{g}\sin 2\pi Nt + \frac{N^2}{N^2-n_s^2}\cdot 4\pi^2 N^2 M_s\sin^2\frac{\pi a_1}{l}\cdot\frac{\delta_0}{g}\sin 2\pi Nt$$

$$+\frac{4F}{\pi}\cdot\frac{N^2}{N^2-n_s^2}\sin\frac{\pi a_2}{l}\sin 2\pi N(t-t_2).$$

The vector components for this central operative force are OA, AC and CD, as shewn in Fig. 43, which only differs from Fig. 39 in that C_1 is coincident with B_1. C is again the middle point of C_1C_2, and the lengths of the various lines are the same as those stated in the construction of Fig. 39. Thus OD is the vector for the central operative force of magnitude

$$2\pi^2 M_G\{(n_0^2-N^2)^2+(2Nn_b)^2\}^{\frac{1}{2}}\frac{\delta_0}{g},$$

and it leads OB_2 by an angle $\tan^{-1}\frac{2Nn_b}{n_0^2-N^2}.$

OA is the vector for

$$-\frac{4\pi^2 N^2}{N^2 - n_s^2}\left[(M_U + M_S)n_s^2 - M_U N^2\right]\left[\sin^2\frac{\pi a_1}{l} + \sin^2\frac{\pi a_2}{l}\right]\frac{\delta_0}{g}\sin 2\pi Nt.$$

$\frac{1}{2}AC_1$ is the vector for

$$\frac{N^2}{N^2 - n_s^2}\cdot 4\pi^2 N^2 M_S \sin^2\frac{\pi a_1}{l}\cdot\frac{\delta_0}{g}\sin 2\pi Nt.$$

$\frac{1}{2}AC_2$ is the vector for

$$\frac{4F}{\pi}\cdot\frac{N^2}{N^2 - n_s^2}\sin\frac{\pi a_2}{l}\sin 2\pi N(t - t_2),$$

C_2 lying on a semi-circle of diameter

$$\frac{2N^2}{N^2 - n_s^2}\cdot 4\pi^2 N^2 M_S \sin^2\frac{\pi a_2}{l}\cdot\frac{\delta_0}{g}.$$

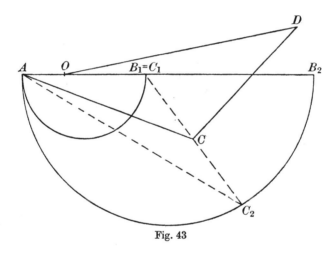

Fig. 43

C being the middle point of $C_1 C_2$, AC is the vector sum of $\frac{1}{2}AC_1$ and $\frac{1}{2}AC_2$. Hence CD, the closing line of the diagram, must be the vector for

$$P\sin(2\pi Nt + \gamma)\left[\sin\frac{\pi a_1}{l} + \sin\frac{\pi a_2}{l}\right].$$

C is located by the condition that the ratio $\dfrac{AC_2}{CD}$ is known.

The scale of the diagram and consequently δ_0 is determined by the fact that the force represented by CD has the known value

$$P\left[\sin\frac{\pi a_1}{l} + \sin\frac{\pi a_2}{l}\right].$$

Figs. 39 and 43 can be combined into the single diagram shewn in Fig. 44.

The dotted curve ACP is the locus of the centres of the vertical intercepts between the circles of which intercepts C_1CC_2 is a specimen. The dotted curve $PC'Q$ is the locus of the middle points of radiating intercepts of which $B_1C'C_2'$ is a specimen. The point C, which has to be located, lies on one or other of these curves, and its exact position is determined by the condition

$$\frac{AC_2}{CD} = \frac{\dfrac{8F}{\pi} \cdot \dfrac{N^2}{N^2 - n_s^2} \sin \dfrac{\pi a_2}{l}}{P\left[\sin \dfrac{\pi a_1}{l} + \sin \dfrac{\pi a_2}{l} \right]}.$$

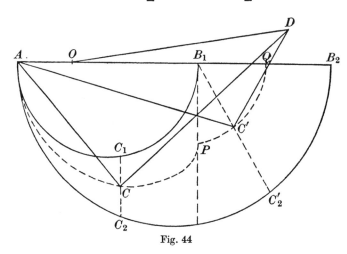

Fig. 44

If this condition places C on the curve ACP, both locomotives are moving on their springs. If C is located on the curve $PC'Q$, Locomotive No. 1 has its springs locked. If both locomotives have locked springs, C is situated at Q.

The foregoing methods will be applied to the case of a bridge of 120 feet span put into oscillation by two skidding locomotives stationed on the bridge 50 feet apart. Two positions of the locomotives will be studied and these are shewn by Figs. 45 (a) and 45 (b).

The two locomotives are identical with their hammer-blows in phase and having a frequency of 5 periods per second. The bridge and locomotive characteristics are as follows:

$$M_G = 200 \text{ tons}, \qquad l = 120 \text{ feet}, \qquad n_0 = 7, \qquad n_b = 0.4.$$
$$M_U = 20 \text{ tons}, \qquad M_S = 60 \text{ tons}, \qquad n_s = 3, \qquad F = 10 \text{ tons},$$
$$P = 15 \text{ tons at 5 revs. per sec.}$$

For Case (1), Fig. 45 (a), $a_1 = 10$ feet, $a_2 = 60$ feet.

$$\sin\frac{\pi a_1}{l} = \sin 15° = 0\cdot 2588, \quad \sin^2\frac{\pi a_1}{l} = 0\cdot 0670,$$

$$\sin\frac{\pi a_2}{l} = \sin 90° = 1, \qquad \sin^2\frac{\pi a_2}{l} = 1,$$

$$OD = 2\pi^2 M_G \{(n_0{}^2 - N^2)^2 + (2Nn_b)^2\}^{\frac{1}{2}} \frac{\delta_0}{g} = 2985\delta_0 \text{ tons,}$$

$$\tan B_2 OD = \frac{2Nn_b}{n_0{}^2 - N^2} = \tfrac{1}{6},$$

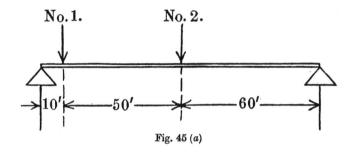

No. 1.   No. 2.

10'   50'   60'

Fig. 45 (a)

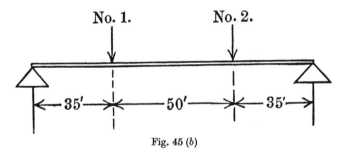

No. 1.   No. 2.

35'   50'   35'

Fig. 45 (b)

$$OA = \frac{4\pi^2 N^2}{N^2 - n_s{}^2}[(M_U + M_S)n_s{}^2 - M_U N^2]\left[\sin^2\frac{\pi a_1}{l} + \sin^2\frac{\pi a_2}{l}\right]\frac{\delta_0}{g} = 450\delta_0 \text{ tons,}$$

$$AB_1 = \frac{2N^2}{N^2 - n_s{}^2}\cdot 4\pi^2 N^2 M_S \sin^2\frac{\pi a_1}{l}\cdot \frac{\delta_0}{g} = 385\delta_0 \text{ tons,}$$

$$AB_2 = \frac{2N^2}{N^2 - n_s{}^2}\cdot 4\pi^2 N^2 M_S \sin^2\frac{\pi a_2}{l}\cdot \frac{\delta_0}{g} = 5751\delta_0 \text{ tons,}$$

$$AC_2 = \frac{8F}{\pi}\cdot \frac{N^2}{N^2 - n_s{}^2}\sin\frac{\pi a_2}{l} = 39\cdot 789 \text{ tons,}$$

$$CD = P\left[\sin\frac{\pi a_1}{l} + \sin\frac{\pi a_2}{l}\right] = 18\cdot 882 \text{ tons,}$$

$$\frac{AC_2}{CD} = 2\cdot 11.$$

Fig. 46 is the vector diagram for this case arranged so that $\dfrac{AC_2}{CD} = 2\cdot 11$, and the position of C shews that the springs are locked for Locomotive No. 1. To the same scale that CD represents $18\cdot 882$ tons, AB_2 scales $48\cdot 254$ tons. Hence $5751\delta_0 = 48\cdot 254$; that is, $\delta_0 = 0\cdot 0084$ foot $= 0\cdot 101$ inch.

For Case (2), Fig. 45 (b), $a_1 = 35$ feet, $a_2 = 85$ feet.

$$\sin\frac{\pi a_1}{l} = \sin\frac{\pi a_2}{l} = \sin 52^\circ\ 30' = 0\cdot 7934,$$

$$\sin^2\frac{\pi a_1}{l} = \sin^2\frac{\pi a_2}{l} = 0\cdot 6294,$$

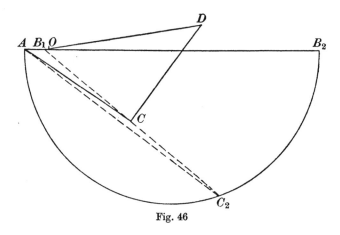

Fig. 46

$$OD = 2\pi^2 M_G\{(n_0{}^2 - N^2)^2 + (2Nn_b)^2\}^{\frac{1}{2}}\frac{\delta_0}{g} = 2985\delta_0 \text{ tons},$$

$$\tan B_2 OD = \frac{2Nn_b}{n_0{}^2 - N^2} = \tfrac{1}{6},$$

$$OA = \frac{4\pi^2 N^2}{N^2 - n_s{}^2}[(M_U + M_S)\,n_s{}^2 - M_U N^2]\left[\sin^2\frac{\pi a_1}{l} + \sin^2\frac{\pi a_2}{l}\right]\frac{\delta_0}{g} = 531\delta_0 \text{ tons},$$

$$AB_1 = AB_2 = \frac{2N^2}{N^2 - n_s{}^2}\cdot 4\pi^2 N^2 M_S \sin^2\frac{\pi a_1}{l}\cdot\frac{\delta_0}{g} = 3619\delta_0 \text{ tons},$$

$$AC_1 = AC_2 = \frac{8F}{\pi}\cdot\frac{N^2}{N^2 - n_s{}^2}\sin\frac{\pi a_1}{l} = 31\cdot 566 \text{ tons},$$

$$CD = P\left[\sin\frac{\pi a_1}{l} + \sin\frac{\pi a_2}{l}\right] = 23\cdot 8005 \text{ tons},$$

$$\frac{AC_2}{CD} = 1\cdot 33.$$

Fig. 47 is the vector diagram for this case, C_2 being located to make $\dfrac{AC_2}{CD} = 1.33$. To the same scale that CD represents 23·8005 tons, AB_2 represents 37·997 tons.

Hence $3619\delta_0 = 37.997$; that is, $\delta_0 = 0.0105$ foot $= 0.126$ inch.

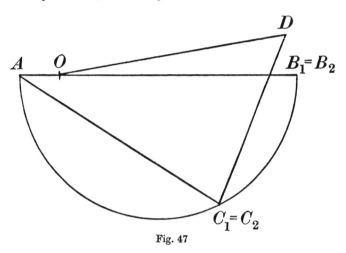

Fig. 47

When a bridge is being oscillated by two skidding locomotives, the magnitude and phase angle of the equivalent central operative force is given by the line OD, but, for calculating the dynamical stresses in the bridge, it is necessary to know the magnitudes and phases of the two operating forces separately, and this can be determined in the following manner. The downward force exerted on the bridge by Locomotive No. 1 is given by

$$R_1 = P \sin\left(2\pi Nt + \gamma\right) + \frac{4F}{\pi} \cdot \frac{N^2}{N^2 - n_s{}^2} \sin 2\pi N\left(t - t_1\right)$$
$$- \frac{4\pi^2 N^2}{N^2 - n_s{}^2}\left[\left(M_U + M_S\right)n_s{}^2 - M_U N^2\right]\frac{\delta_0}{g}\sin\frac{\pi a_1}{l}\sin 2\pi Nt,$$

and the corresponding vector diagram, such as Fig. 40, gives the values of δ_0, γ and $2\pi n t_1$, which enter into the composition of R_1. Hence R_1 can be viewed as the sum of the projections of three vectors OF, FG_1, G_1H_1, see Fig. 48, where

OF, the vector for $P\sin\left(2\pi Nt + \gamma\right)$,

is parallel to CD,

FG_1, the vector for $\dfrac{4F}{\pi} \cdot \dfrac{N^2}{N^2 - n_s{}^2}\sin 2\pi N\left(t - t_1\right)$,

is parallel to AC_1, and

G_1H_1, the vector for $\dfrac{4\pi^2 N^2}{N^2 - n_s{}^2}\left[\left(M_U + M_S\right)n_s{}^2 - M_U N^2\right]\dfrac{\delta_0}{g}\sin\dfrac{\pi a_1}{l}\sin 2\pi Nt$,

is parallel to OA.

Taking the particular case illustrated by Fig. 40,

OF represents 15 tons,
FG_1 „   19·89 tons,
G_1H_1 „   3·10 tons.

Similarly the other operative force can be viewed as the sum of the projections of three vectors, OF, FG_2, G_2H_2, where

OF represents 15 tons and is parallel to CD,
FG_2 „   19·89 tons   „ AC_2,
G_2H_2 „   3·80 tons   „ OA.

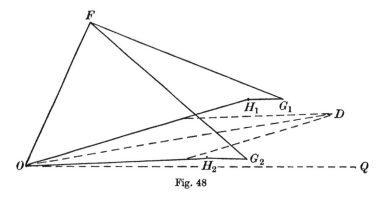

Fig. 48

In Fig. 48 these vectors are set out, and it appears that OH_1, which is the vector for R_1, the downward force produced by Locomotive No. 1, has a magnitude of 22·8 tons, and the magnitude of R_2, as given by the vector OH_2, has a magnitude of 18·0 tons. The equivalent central operative force

$$R_1 \sin\frac{\pi a_1}{l} + R_2 \sin\frac{\pi a_2}{l}$$

is given by the vector OD on the figure. This scales off to be 31 tons, and in magnitude and direction is in agreement with the vector OD on Fig. 40.

If a locomotive, say No. 1, has its springs locked, the downward force R_1 it exerts on the bridge is given by the simplified expression

$$R_1 = P \sin(2\pi Nt + \gamma) + 4\pi^2 N^2 (M_U + M_S)\frac{\delta_0}{g}\sin\frac{\pi a_1}{l}\sin 2\pi Nt.$$

These methods for determining operative forces will be utilized in a later Chapter for calculating the dynamical bending-moments and shearing-forces induced in a bridge by the hammer-blows of locomotives.

CHAPTER XII

OSCILLATIONS IN SHORT-SPAN BRIDGES

The method for predicting bridge oscillations by means of a stationary skidding locomotive, developed in the previous Chapters, has limitations which preclude its application to short-span bridges. For bridges of less than about 50 feet span, the time during which a locomotive travelling at high speed occupies the central region of the bridge is insufficient to allow the oscillations to mount up to the full value given by the corresponding skidding locomotive; furthermore, the idealization of a locomotive into concentrated sprung and unsprung masses is hardly legitimate, when dealing with a bridge whose length may be less than the wheel-base of the locomotive. Hence, for short-span bridges, other methods for determining the oscillations must be sought, and, fortunately, theory confirmed by experiment finds a solution by the following pleasingly simple process.

Treat the hammer-blows associated with the various axles of the locomotive as though they were statical forces. The variation in the deflections, worked out on this assumption, will be found to agree very closely in magnitude with the true state of oscillation, though the latter will lag behind the former to a considerable extent owing to the heavy damping associated with short-span bridges.

That the operative forces, acting on a short-span bridge, should be statical in their effect is obvious, since the dynamic magnification is

$$\frac{n_0{}^2}{\{(n_0{}^2 - N^2)^2 + (2Nn_b)^2\}^{\frac{1}{2}}},$$

and, when n_0 is large compared with N, as it is for short spans, this expression tends towards unity.

It is not, however, immediately obvious that the operative forces can be treated as having the same magnitude as the hammer-blows, since, at first sight, this is apparently equivalent to neglecting the inertia effects of the locomotive and the effects of spring friction.

Actually, the equality in magnitude between the operative forces and the hammer-blows is mainly brought about by the heavy damping in short spans and, in consequence, the component force which has to be added to a hammer-blow to give the corresponding operative force produces a resultant which, though out of phase with the hammer-blow, hardly differs from it in magnitude.

This point will be illustrated by the vector diagrams for two bridges of 60 and 30 feet span, acted on by the same central bridge oscillator.

For the oscillator $M_U = 20$ tons, $M_S = 60$ tons, $n_s = 3$, $F = 8$ tons, N, the speed of rotation, is 6 per second, P at this speed is 21·6 tons.

For Bridge No. 1. $l = 60$ feet, $M_G = 100$ tons, $n_0^2 = 100$, $n_b = 2.5$. This value of n_b gives a ratio of successive residual deflections of 0·2.

For Bridge No. 2. $l = 30$ feet, $M_G = 40$ tons, $n_0^2 = 200$, $n_b = 6.0$. This value of n_b gives a ratio of successive residual deflections of 0·07, the oscillations being almost dead-beat in character.

For Bridge No. 1. If δ_0 is expressed in feet,

$$OD = 2\pi^2 M_G \{(n_0^2 - N^2)^2 + (2Nn_b)^2\}^{\frac{1}{2}} \frac{\delta_0}{g} \text{ tons} = 4336\delta_0 \text{ tons},$$

$$\tan BOD = \frac{2Nn_b}{n_0^2 - N^2} = \frac{15}{32},$$

$$OA = \frac{4\pi^2 N^2}{N^2 - n_s^2} [(M_U + M_S) n_s^2 - M_U N^2] \frac{\delta_0}{g} \text{ tons} = 0,$$

$$AB = \frac{N^2}{N^2 - n_s^2} \cdot 4\pi^2 N^2 M_S \frac{\delta_0}{g} \text{ tons} = 3542\delta_0 \text{ tons},$$

$$AC = \frac{4F}{\pi} \cdot \frac{N^2}{N^2 - n_s^2} \text{ tons} = 13.5812 \text{ tons},$$

$$CD = P = 21.6 \text{ tons},$$

$$\frac{AC}{CD} = 0.629.$$

The statical deflection $D_P = \dfrac{Pg}{2\pi^2 n_0^2 M_G}$ foot $= 0.0035$ foot.

For Bridge No. 2, the corresponding dimensions are:

$$OD = 4395\delta_0 \text{ tons}, \quad \tan BOD = \frac{18}{41}, \quad OA = 0, \quad AB = 3542\delta_0 \text{ tons},$$

$$AC = 13.5812 \text{ tons}, \quad CD = 21.6 \text{ tons}, \quad \frac{AC}{CD} = 0.629,$$

and
$$D_P = 0.00440 \text{ foot}.$$

The two corresponding vector diagrams are shewn by Figs. 49 and 50.

For Fig. 49, $l = 60$ feet, the operative force OD scales 22·62 tons, and $\delta_0 = 0.0052$ foot. Hence the dynamic magnification $= \dfrac{0.0052}{0.0035} = 1.48$.

For Fig. 50, $l = 30$ feet, the operative force OD scales 22·93 tons, and $\delta_0 = 0.0052$ foot. Hence the dynamic magnification $= \dfrac{0.0052}{0.0044} = 1.18$.

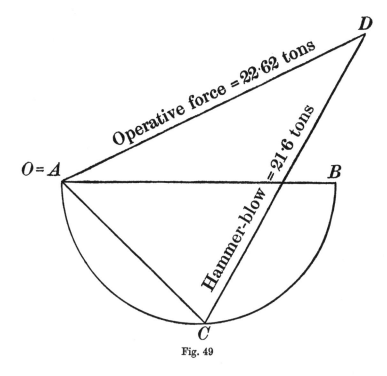

Fig. 49

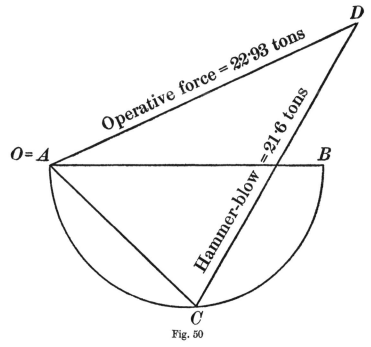

Fig. 50

In both cases, it will be seen that the operative force differs but little in magnitude from the hammer-blow and that the oscillation lags behind the hammer-blow by a phase angle of about 60°. For the shorter span, its comparatively high natural frequency of 14·14 has the effect of reducing the dynamic magnification almost to unity, and this example gives a theoretical justification for ignoring dynamical effects in short-span bridges and obtaining deflections due to hammer-blows by purely statical methods.

An examination of the deflection records for short-span bridges given in the Bridge Stress Report confirms this theoretical conclusion. Fig. 51 is one of these records, and refers to a bridge of 33 feet span having a total mass of 34·3 tons and an unloaded frequency of 14. The frequency of the hammer-blows was 6·48 periods per second. As is always the case for very short-span

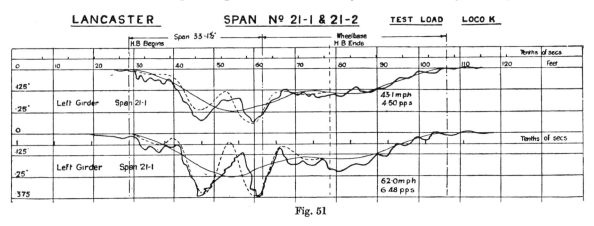

Fig. 51

bridges, the smoothness of the deflection record is marred by many irregularities, but nevertheless it will be seen that the maximum downward deflection given by calculation and shewn dotted in the diagram is in good agreement with that recorded by experiment, and for the oscillation which occurs when the "crawl-deflection" is a maximum the phase difference between the oscillation and the hammer-blow is of the order of 60°, which again is in accordance with theory.

On the assumption that hammer-blows can be treated as statical forces, the most convenient method for obtaining the corresponding deflection diagram is as follows.

Suppose the locomotive has four coupled axles, 1, 2, 3 and 4, the distances of 2, 3 and 4 from 1 being a_2, a_3 and a_4.

Suppose the hammer-blows (measured in tons) associated with the four axles are

$$P_1 \sin 2\pi Nt, \quad P_2 \sin (2\pi Nt + \phi_2), \quad P_3 \sin (2\pi Nt + \phi_3) \quad \text{and} \quad P_4 \sin (2\pi Nt + \phi_4),$$

the revolutions of the driving-wheels being N per second. When the leading axle, No. 1, has advanced a distance x along the bridge and consequently the other axles have advanced distances $x - a_2$, $x - a_3$ and $x - a_4$, the central deflection for the combined action of the hammer-blows is

$$D_1 \left[P_1 \sin \frac{\pi x}{l} \sin 2\pi N t + P_2 \sin \frac{\pi (x - a_2)}{l} \sin (2\pi N t + \phi_2) \right.$$

$$\left. + P_3 \sin \frac{\pi (x - a_3)}{l} \sin (2\pi N t + \phi_3) + P_4 \sin \frac{\pi (x - a_4)}{l} \sin (2\pi N t + \phi_4) \right],$$

where D_1 is the central deflection due to a central force of one ton, that is,

$$D_1 = \frac{g}{2\pi^2 n_0{}^2 M_G} \text{ feet, if } M_G \text{ is measured in tons.}$$

This expression for the central deflection can be written in the form

$$D_1 \left[\left(A \sin \frac{\pi x}{l} - B \cos \frac{\pi x}{l} \right) \sin 2\pi N t + \left(C \sin \frac{\pi x}{l} - D \cos \frac{\pi x}{l} \right) \cos 2\pi N t \right],$$

where $A = P_1 + P_2 \cos \frac{\pi a_2}{l} \cos \phi_2 + P_3 \cos \frac{\pi a_3}{l} \cos \phi_3 + P_4 \cos \frac{\pi a_4}{l} \cos \phi_4,$

$$B = \qquad P_2 \sin \frac{\pi a_2}{l} \cos \phi_2 + P_3 \sin \frac{\pi a_3}{l} \cos \phi_3 + P_4 \sin \frac{\pi a_4}{l} \cos \phi_4,$$

$$C = \qquad P_2 \cos \frac{\pi a_2}{l} \sin \phi_2 + P_3 \cos \frac{\pi a_3}{l} \sin \phi_3 + P_4 \cos \frac{\pi a_4}{l} \sin \phi_4,$$

$$D = \qquad P_2 \sin \frac{\pi a_2}{l} \sin \phi_2 + P_3 \sin \frac{\pi a_3}{l} \sin \phi_3 + P_4 \sin \frac{\pi a_4}{l} \sin \phi_4.$$

For any position of the locomotive, defined by x, the semi-amplitude of the oscillation has the magnitude

$$D_1 \left\{ \left(A \sin \frac{\pi x}{l} - B \cos \frac{\pi x}{l} \right)^2 + \left(C \sin \frac{\pi x}{l} - D \cos \frac{\pi x}{l} \right)^2 \right\}^{\frac{1}{2}},$$

that is,

$$D_1 \left\{ \tfrac{1}{2} (A^2 + B^2 + C^2 + D^2) - \tfrac{1}{2} (A^2 + C^2 - B^2 - D^2) \cos \frac{2\pi x}{l} \right.$$

$$\left. - (AB + CD) \sin \frac{2\pi x}{l} \right\}^{\frac{1}{2}}.$$

Accordingly, if a defines the position of the axle No. 1, when the locomotive is giving the maximum "crawl-deflection", the increment of central deflection due to hammer-blow is

$$D_1 \left\{ \tfrac{1}{2} (A^2 + B^2 + C^2 + D^2) - \tfrac{1}{2} (A^2 + C^2 - B^2 - D^2) \cos \frac{2\pi a}{l} \right.$$

$$\left. - (AB + CD) \sin \frac{2\pi a}{l} \right\}^{\frac{1}{2}}.$$

The most convenient method of obtaining the "crawl-deflection" is as follows. Suppose W_1, W_2, W_3, ... are the axle loads on the bridge and their distances from one end are x_1, x_2, x_3, ... The corresponding central deflection is

$$D_1\left[W_1\sin\frac{\pi x_1}{l}+W_2\sin\frac{\pi x_2}{l}+W_3\sin\frac{\pi x_3}{l}+...\right].$$

This method of determining dynamical deflections for short spans will now be illustrated by two particular cases.

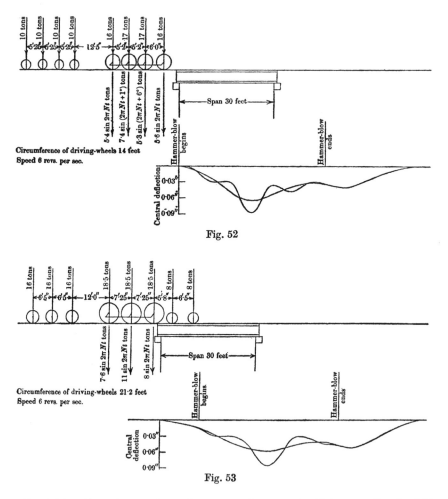

Fig. 52

Fig. 53

The bridge in each case is 30 feet span; its mass is 40 tons and $n_0{}^2 = 200$. Hence the central deflection D_1, produced by a steady central force of 1 ton, as given by $D_1 = \dfrac{g}{2\pi^2 n_0{}^2 M_G}$, makes $D_1 = 0{\cdot}000204$ foot. The locomotive characteristics are shewn on Figs. 52 and 53.

For the eight-coupled locomotive, when all four driving axles are on the bridge and the leading driving axle has advanced a distance x feet along the span, the central dynamical increment of deflection, measured in feet, has the value

$$\left(0 \cdot 002436 \sin\frac{\pi x}{l} - 0 \cdot 003113 \cos\frac{\pi x}{l}\right) \sin 2\pi Nt$$
$$- \left(0 \cdot 000102 \sin\frac{\pi x}{l} - 0 \cdot 000091 \cos\frac{\pi x}{l}\right) \cos 2\pi Nt.$$

The greatest value of this is 0·00395 foot, and occurs when $x = 23$ feet approximately. The "crawl-deflection" for this particular position of the load is 0·01093 foot.

For the six-coupled locomotive, the corresponding general expression for the dynamic increment of the central deflection is

$$\left(0 \cdot 003341 \sin\frac{\pi x}{l} - 0 \cdot 003093 \cos\frac{\pi x}{l}\right) \sin 2\pi Nt.$$

The greatest value of this is 0·00455 foot, and occurs when $x = 22 \cdot 5$ feet approximately. The "crawl-deflection" for this particular position of the load is 0·00958 foot.

The "crawl-deflections" and the deflections as augmented by hammer-blow are shewn for complete runs on Figs. 52 and 53.

CHAPTER XIII

BRIDGE AND LOCOMOTIVE CHARACTERISTICS

The validity of predictions made by mathematical analysis depends upon the accuracy with which numerical values can be assigned to the various bridge and locomotive characteristics entering into the composition of the formulae deduced.

For predicting oscillations in railway bridges, the bridge characteristics which must be known are

(1) The length of the bridge, l.

(2) The total mass, M_G.

(3) The unloaded fundamental frequency, n_0.

(4) The damping coefficient, n_b.

The locomotive characteristics required are

(5) The hammer-blow at 1 rev. per sec.

(6) The circumference of the driving-wheels.

(7) The spring-borne mass, M_S ⎫

(8) The unsprung mass, M_U ⎬ excluding the tender.

(9) The natural frequency of the locomotive on its springs, n_s.

(10) The frictional force resisting spring movements, F.

For long-span bridges, in which spring movement does not occur, n_s and F are not required, and furthermore, for short-span bridges, it is not necessary to know the value of n_b.

Bridge characteristics

If the central deflection D_W feet produced by a steady central load W tons is known, the unloaded frequency of the bridge n_0 can be deduced by the formula

$$2\pi^2 n_0^2 = \frac{Wg}{D_W M_G},$$

where M_G, the total mass of the bridge, is measured in tons.

Alternatively, if I, the moment of inertia of the mid-cross-section, is known, n_0 can be determined by the formula

$$4\pi^2 n_0^2 = \frac{\pi^4}{l^3} \cdot \frac{EIg}{M_G},$$

where I and l are measured in feet units, M_G in tons, and E in tons per square foot.

In the Report of the Bridge Stress Committee, curves are shewn indicating how the values of M_G and n_0 varied with the span for the particular bridges which were tested. For bridges of the same span these characteristics exhibit wide variations, but for double-track spans ranging from 40 to 300 feet, reasonable mean values are given by the formulae

$$M_G = 2 \cdot 4l + \frac{1}{320} l^2 \text{ tons,}$$

and
$$n_0 = 4 \cdot 86 - 0 \cdot 0103l + \frac{222}{l},$$

where l is the span measured in feet.

The values of M_G and n_0 thus deduced and also the values of D_1, the central deflection produced by a steady central load of one ton, as given by the formula $D_1 = \dfrac{12g}{2\pi^2 n_0{}^2 M_G}$ ins., are tabulated below, and curves shewing how these mean values of M_G and n_0 vary with the span are given by Figs. 54 and 55.

For single-track bridges the frequencies will be much the same, but the masses will be halved and the values of D_1 will be doubled approximately.

Mean values of mass and frequency for double-track bridges

Span in feet	M_G in tons	n_0	D_1 in ins.
40	100	10·00	$1 \cdot 94 \times 10^{-3}$
60	155	7·94	2·00 ,,
80	212	6·81	1·99 ,,
100	271	6·05	1·97 ,,
120	333	5·48	1·96 ,,
140	397	5·00	1·96 ,,
160	464	4·60	1·99 ,,
180	533	4·24	2·04 ,,
200	608	3·91	2·10 ,,
220	679	3·60	2·22 ,,
240	756	3·31	2·36 ,,
260	835	3·04	2·54 ,,
280	917	2·77	2·78 ,,
300	1000	2·51	3·10 ,,

It will be seen from the table that D_1 is almost constant for spans varying from 40 to 200 feet, and this means that within these limits $n_0{}^2$ varies inversely as M_G. This relationship follows from the fact that, for moderate or short spans, the weight of the longitudinal supporting girders contributes but little to the total stress for which these girders are designed, and it can be demonstrated as follows:

Let w_l per unit length be the equivalent uniformly distributed load the bridges have to carry, in addition to their dead load.

Let w_d per unit length be the weight of decking and track which will be independent of the span.

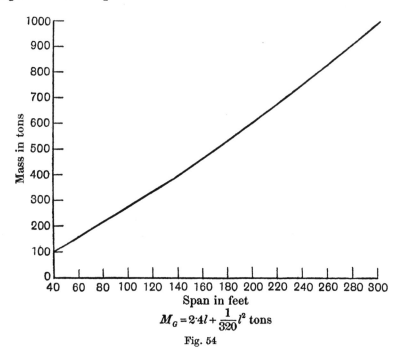

$$M_G = 2{\cdot}4l + \frac{1}{320}l^2 \text{ tons}$$

Fig. 54

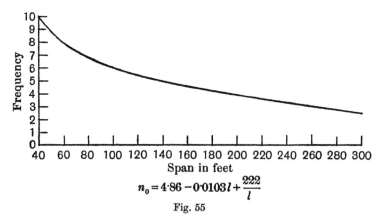

$$n_0 = 4{\cdot}86 - 0{\cdot}0103\,l + \frac{222}{l}$$

Fig. 55

The central bending-moment due to the combined action of these loads is $(w_l+w_d)\dfrac{l^2}{8}$ and the corresponding flange-stress is $\dfrac{(w_l+w_d)\,l^2 d}{16I}$, where d is the semi-depth of the supporting girders (assumed to be two in number).

On the assumption that d is proportional to l and that stresses due to the weight of the longitudinal main girders are negligible, it follows that for equality of flange-stress $\dfrac{l^3}{I}$ must be constant and, since the central deflection due to a given central load is proportional to $\dfrac{l^3}{I}$, this central deflection will be independent of the span.

Hence, if bridges in which the ratio of depth to span is constant are designed for a given maximum flange-stress, then, on the assumption that the equivalent live load and the decking load are the same per unit length and that the stresses due to the weight of the main longitudinal girders are relatively insignificant, it follows that the central deflection produced by a given central load is independent of the span.

This condition, in its turn, implies that $n_0^2 M_G$ is a constant, and from the results tabulated above it appears that this simple connection between two important bridge characteristics is satisfied approximately for spans ranging from 40 to 200 feet, but beyond this upper limit it breaks down owing to the increasing importance of the stresses set up by the weight of the main longitudinal girders.

Within this range, $l = 40$ feet to $l = 200$ feet, $n_0^2 M_G = 10,000$ approximately, for average weight double-track bridges, M_G being expressed in tons.

For average weight single-track bridges the corresponding relation is $n_0^2 M_G = 5,000$.

BRIDGE DAMPING

This action is made apparent by the more or less rapid subsidence of the oscillations after a locomotive has passed across a bridge. Damping is brought about by imperfect elasticity in the structure and in the track which it supports, and it may be augmented to some extent by friction at the end bearings and the dissipation of energy consequent on any lack of rigidity in the abutments. For long-span bridges the damping is so small that residual oscillations may persist for a considerable period, but for short spans it is so intense that free oscillations are almost dead-beat in character and residual oscillations can hardly be detected.

In all the foregoing analysis bridge damping has been taken into account by treating it as a resistance to movement distributed along the span, this resistance to movement being written in the form $4\pi n_b m v$ per unit length, where v is the vertical velocity of the girder at the section under consideration, m is the mass of the bridge per unit length, and n_b is a coefficient which defines the damping.

Taking damping into account in this manner, the ratio of successive residual deflections is given by $e^{-2\pi\frac{nb}{n_0}}$, and, from observations of this ratio, the value of n_b appropriate to any particular bridge can be deduced.

In addition to a resistance distributed along the span, there is another damping influence due to the continuity of the track at the abutments, which sets up couples of a non-elastic character resisting angular deflections at the ends of a bridge. In long-span bridges this particular damping effect is unimportant, but in short-span bridges it is relatively large and is the cause of the heavy damping invariably found in bridges of this class.

Damping due to terminal couples can be taken into account in the following manner.

Suppose the residual oscillation of the bridge is given by $y = f(t)\sin\frac{\pi x}{l}$. If the damping resistance per unit length distributed along the span is expressed in the form $K\frac{df}{dt}\sin\frac{\pi x}{l}$, then the bending-moment thus produced at any section is

$$\frac{l^2}{\pi^2} K \frac{df}{dt}\sin\frac{\pi x}{l}.$$

The terminal damping couples due to the continuity of the track are assumed to be proportional to the rate of change of angular deflection at the ends of the bridge, and can be written in the form $\mu\frac{\pi}{l}\frac{df}{dt}$, since $\frac{\pi}{l}\frac{df}{dt}$ is the rate of change of the end slope.

Hence the combined bending-moment due to the two types of damping resistance is

$$\left(\frac{l^2}{\pi^2} K \sin\frac{\pi x}{l} + \mu\frac{\pi}{l}\right)\frac{df}{dt}.$$

For deducing deflections no appreciable error will be introduced if the uniformly distributed bending-moment $\mu\frac{\pi}{l}\frac{df}{dt}$ is replaced by its primary harmonic component $\frac{4\mu}{l}\frac{df}{dt}\sin\frac{\pi x}{l}$. The combined bending-moment then takes the form

$$\frac{l^2}{\pi^2}\left(K + \mu\frac{4\pi^2}{l^3}\right)\frac{df}{dt}\sin\frac{\pi x}{l},$$

and this is the same as the bending-moment produced by a damping resistance $\left(K + \mu\frac{4\pi^2}{l^3}\right)v$ per unit length of span.

Writing this damping resistance in the usual form $4\pi n_b mv$, it appears that

$$n_b = \frac{1}{4\pi m}\left(K + \mu \frac{4\pi^2}{l^3}\right),$$

and a general expression for n_b is obtained in the form

$$n_b = \frac{l}{M_G}\left(a + \frac{b}{l^3}\right).$$

For long spans, the second term in the bracket is small compared with the first, but, for short spans, it is the dominating contribution, and this general formula explains the fact that short-span bridges, say of 50 feet span or less, are always found to be very heavily damped.

For bridges which have the average masses and frequencies previously tabulated, the corresponding values of n_b can be obtained from the formula

$$n_b = \frac{l}{M_G}\left(0{\cdot}12 + \frac{0{\cdot}63 \times 10^6}{l^3}\right).$$

For spans ranging from 80 to 300 feet, values of n_b thus deduced are in good agreement with those employed for the calculations in the Report of the Bridge Stress Committee. For short spans this formula takes into account the very marked increase in damping associated with bridges of this class, and the higher values of n_b it prescribes, for spans less than 80 feet, are more in accordance with reality than those adopted in the Report of the Bridge Stress Committee.

The values of n_b given by the above formula and the ratio of successive residual deflections, as given by the formula $\rho = e^{-\frac{2\pi n_b}{n_0}}$, are tabulated below, and the result of plotting n_b and ρ in terms of l are shewn on Figs. 56 and 57.

Values of the damping coefficient for double-track
bridges of average mass

Span in feet	n_b	$\rho = e^{-\frac{2\pi n_b}{n_0}}$
40	3·986	0·082
60	1·176	0·394
80	0·510	0·625
100	0·277	0·750
120	0·175	0·819
140	0·123	0·857
160	0·094	0·879
180	0·077	0·892
200	0·065	0·900
220	0·058	0·904
240	0·053	0·905
260	0·049	0·905
280	0·045	0·902
300	0·043	0·898

For single-track bridges, the values of n_b tabulated above still apply, since the reduction in the damping resistance per unit length is accompanied by an almost equal reduction in the mass of the bridge.

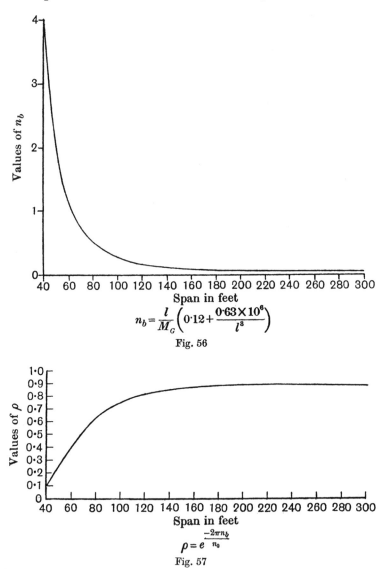

$$n_b = \frac{l}{M_G}\left(0\cdot12 + \frac{0\cdot63 \times 10^6}{l^3}\right)$$

Fig. 56

$$\rho = e^{\frac{-2\pi n_b}{n_0}}$$

Fig. 57

For any given span the characteristics M_G, n_0 and n_b vary between wide limits in accordance with variations in design, and formulae expressing these characteristics in terms of span can, at the best, only be regarded as giving an indication of possible values. In particular, information about the bridge characteristic n_b is disappointingly meagre, and before it will be possible to assign precise numerical values to this coefficient applicable to all types of railway bridges much more experimental evidence must be made available. It can be obtained by observing the rate of decrease of residual oscillations, but this method may possibly lead to an under-statement, since it is probable that the damping increases somewhat when the bridge is loaded. Most certainly it is impossible to express the bridge characteristics M_a, n_0 and n_d as a function of span only and consequently any attempt to lay down impact allowances in terms of span is a hopeless quest. It could only be done if all bridges were standardized to conform to one single type. General formulae for impact allowances embodying all the essential bridge and locomotive characteristics can, however, be evolved and this is done in the next Chapter; but, for any particular case, the characteristics appropriate to that case must be employed and these, as in some of the numerical cases which have been worked out, may differ widely from the values tabulated above. For any particular railway system the determination of the bridge and locomotive characteristics limited by that system would not be a very formidable task. From the information so obtained, and in that way only, can the characteristics and impact allowances applicable to future construction be predicted in a thoroughly scientific and reliable manner.

LOCOMOTIVE CHARACTERISTICS
HAMMER-BLOW

This action is due to the balance-weights attached to the wheels of a locomotive for the purpose of minimizing the horizontal inertia forces set up by its reciprocating masses.

The intensity of the hammer-blow varies with the square of the speed of revolution of the wheels, and its magnitude P_1 at one revolution per second is a locomotive characteristic about which no element of uncertainty need exist.

Although a detailed analysis of the inertia forces due to the moving parts of a locomotive would overburden this Chapter, a few results (without proof) are now given to indicate generally how hammer-blows are generated and how they can be computed.

Notation.

For one set of moving parts:

Let M_p denote the mass of the piston, piston-rod and crosshead.

Let M_c denote the mass of the connecting-rod, and l its length.

Let the c.g. of the connecting-rod be distant a from the small end and distant b from the big end.

Let k denote the radius of gyration of the connecting-rod about an axis
through its C.G., perpendicular to the plane of motion.

Let r be the length of the crank.

Let $\omega = 2\pi N$ be the angular velocity of the crank.

Let α be the acceleration of the piston.

Let Ω be the angular acceleration of the connecting-rod.

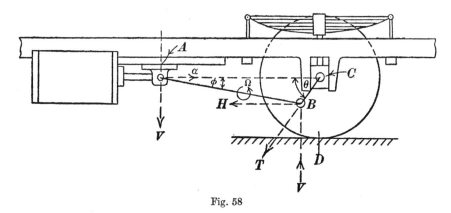

Fig. 58

Fig. 58 is a diagrammatic representation of the mechanism operated by
one cylinder, and the forces acting on the slide-bar at A and on the crank-
pin B, due to the inertia of the piston and connecting-rod, are indicated by
H, T and V, where

$$H = \left(M_P + \frac{b}{l} M_C \right) \alpha,$$

$$T = \frac{a}{l} M_C r \omega^2,$$

$$V = \left(M_P + \frac{b}{l} M_C \right) \alpha \tan \phi + \frac{ab - k^2}{l \cos \phi} M_C \frac{d\Omega}{dt}.$$

The force T which acts outwards along the crank is of constant magnitude
and, consequently, it can be completely neutralized by a suitable counter-
weight. Assuming that this is done, the inertia force exerted by the axle-box
on the frame of the engine in the direction of the line of stroke reduces to

$$H - \frac{2r}{D} (H \sin \theta + V \cos \theta),$$

where D is the diameter of the driving-wheel, and the problem of balancing
a locomotive resolves itself into ways and means of neutralizing or partially
neutralizing this inertia force by means of balance-weights attached to the
driving-wheels.

It is evident that by this means only the primary component of this force can be neutralized, and this primary component is found to have the value

$$\left(M_P+\frac{b}{l}M_C\right)\left[\cos\theta+\frac{r}{D}\left(\frac{R}{2}+\frac{R^3}{8}+\frac{15R^6}{256}+\dots\right)\sin\theta\right]r\omega^2,$$

where R denotes the ratio $\frac{r}{l}$.

In a locomotive $\frac{rR}{2D}$ is a fraction whose value will hardly exceed $\frac{1}{100}$; consequently, the term associated with $\sin\theta$ is quite negligible, and to a sufficient degree of accuracy the primary component of the inertia force on the frame of the engine acting along the line of stroke may be taken to have the value

$$\left(M_P+\frac{b}{l}M_C\right)r\omega^2\cos\theta.$$

If the balance-weights attached to the driving-wheels—over and above those required to balance crank-pins, side rods and the constant force $\frac{a}{l}M_C r\omega^2$ along CB—produce a resultant force whose horizontal and vertical components are $Mr\omega^2\cos\theta$ and $Mr\omega^2\sin\theta$ respectively, the primary component of the inertia force on the locomotive along the line of stroke is reduced to

$$\left(M_P+\frac{b}{l}M_C-M\right)r\omega^2\cos\theta,$$

but this is only achieved at the cost of introducing a fluctuation in rail pressure of magnitude $Mr\omega^2\sin\theta$, which is termed "Hammer-Blow".

The two vertical inertia forces V shewn on Fig. 58 being equal and opposite produce no change in the total downward force exerted by the locomotive, though the varying position of the force V at the slide-bar will cause alterations in the distribution of the total load among the wheels, and a similar variation in the load distribution is also produced by the action of steam pressure in the cylinder. The vertical reaction at the slide-bars acts on the spring-borne part of the locomotive, and if the springs were very free this would produce variations of rail pressure consequent on spring movement. The magnitude of V is, however, so small and the friction in the spring movement of a locomotive is so great that possible fluctuations of total rail pressure due to this cause need not be taken into account.

If the locomotive has two cylinders with cranks at right angles, the combined hammer-blow is

$$\sqrt{2}Mr\,.\,4\pi^2N^2\sin\theta,$$

where N denotes the revs. per sec. of the driving-wheels.

The resultant longitudinal inertia force is

$$\sqrt{2}\left(M_P + \frac{b}{l}M_C - M\right)r \, . \, 4\pi^2 N^2 \cos\theta,$$

and, if d is the distance between the two lines of stroke, the two longitudinal inertia forces will produce a couple of magnitude

$$\left(M_P + \frac{b}{l}M_C - M\right)\frac{rd}{\sqrt{2}} \, . \, 4\pi^2 N^2 \cos\left(\theta + \frac{\pi}{4}\right),$$

tending to swerve the locomotive across the track.

In the case of two-cylinder locomotives a complete balance of the horizontal inertia forces would entail such heavy hammer-blows that only a proportion ranging from 33 to 60 per cent. is neutralized. It is usually stated that some such neutralization is essential, otherwise the locomotive will be unsteady in its running and develop a tendency to swerve from side to side. Such swerving is often noticeable, but its frequency shews no connection with that of the inertia couple and corresponds more with the deviation due to the coning of the treads on the driving-wheels. Experimentally, there seems to be little or no evidence of swerve arising from inertia couples, and the following numerical calculation suggests that the deviating effect of the inertia swerving couple is so small that it could hardly be observed.

For a two-cylinder locomotive without any balance for the reciprocating parts, the primary inertia swerving couple at 5 revs. per sec. will be of the order of 50 tons-feet at its maximum.

If the locomotive is pivoted so that it is perfectly free to make angular movements about a vertical axis through its c.g., then taking its total mass (excluding tender) to be 80 tons and its radius of gyration about a vertical axis through its c.g. as 7 feet, the equation giving θ, its angular movement when the swerving couple alternates 5 times per sec., is

$$80 \times 49 \frac{d^2\theta}{dt^2} = 50g \cos 10\pi t.$$

If the motion starts from a condition of zero angular displacement and zero velocity, the solution of this equation is $\theta = \frac{1}{2500}(1 - \cos 10\pi t)$ approximately, and the maximum angular movement is $\frac{1}{1250} = 2 \cdot 7$ minutes. Bearing in mind that the locomotive is by no means freely pivoted and deviations are resisted by friction, it is difficult to see how any noticeable swerve can be directly attributed to inertia couples.

For the same case the unbalanced longitudinal inertia force would be of the order $20 \cos 10\pi t$ tons and, if the locomotive was suspended so as to be perfectly free to perform longitudinal oscillations, its deviation from the mean position would only be about $\frac{1}{10}$ inch, a quantity so small that it could hardly produce "snatching" at the draw-bar unless some form of resonance was established.

Although unbalanced inertia forces may not be the real cause of swerving, it is quite likely that their existence has an injurious effect on the wear of bearings and in a general straining of the framework. But in so far as cases are on record where locomotives have operated successfully without any balance for their reciprocating parts and practice varies so widely in the amount which is actually balanced, it would appear that finality has not been reached in this problem of locomotive design.

From the point of view of wear of rails and tyres and the reduction of dynamical stresses in bridges, it is obviously desirable that the weights attached to driving-wheels for the purpose of minimizing longitudinal inertia forces should be the least possible consistent with the smooth running and economical maintenance of a locomotive.

For three-cylinder and four-cylinder locomotives, the primary longitudinal inertia forces automatically combine to give a zero resultant. Balance-weights of comparatively small magnitude are generally introduced to eliminate the primary swerving couple due to inertia, but the hammer-blows thus developed seldom exceed 2 tons at 5 revs. per sec., whereas, for two-cylinder locomotives, hammer-blows of 12 tons and upwards are quite usual for this same speed.

Spring frequency

To take into account the effects of spring movement, the locomotive was idealized into a mass M_S, spring supported on a single axle, as shewn in Fig. 29, Chapter IX. The legitimacy of this idealization will now be considered.

If the locomotive springs are such that, when an additional load is applied at the c.g. of the spring-borne mass, the body of the locomotive descends without "fore and aft" tilt, vertical oscillations are quite independent of angular movements, and the idealization mentioned above is justified. The condition for this is that, if $\mu_1, \mu_2, \mu_3, \ldots$ etc. are the forces required to give unit deflections to the various axle-springs, and if $W_1, W_2, W_3, \ldots$ etc. are the corresponding statical axle loads, then $\dfrac{\mu_1}{W_1} = \dfrac{\mu_2}{W_2} = \dfrac{\mu_3}{W_3}$, etc. This condition is generally satisfied, at any rate approximately, but, if it is not, a vertical load applied to the c.g. of the spring-borne part of the locomotive will not deflect all the springs equally and the body of the locomotive will alter its "fore and aft" trim. Consequently, vertical oscillations of the track will produce vertical oscillations of the body of the locomotive, combined with angular movements in a "fore and aft" direction. How far these latter may be of importance will now be studied with reference to a six-wheeled locomotive diagrammatically represented by Fig. 59, in which the strengths of the springs are different and are defined by the quantities μ_1, μ_2 and μ_3.

Let M_s be the spring-borne mass of the locomotive, and $M_s k^2$ its moment of inertia about the transverse horizontal axis through its c.g.

Let $\delta = \delta_0 \sin 2\pi Nt$ be the vertical oscillation imposed on its wheels by the track.

Let z be the downward displacement of the c.g. of M_s below its mean level.

Let θ be the angular tilt of M_s from the horizontal.

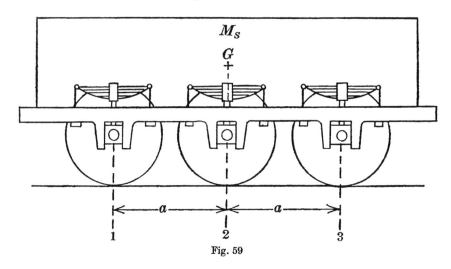

Fig. 59

The increased spring deflections for axles 1, 2 and 3 are

$$z - a\theta - \delta_0 \sin 2\pi Nt, \quad z - \delta_0 \sin 2\pi Nt, \quad z + a\theta - \delta_0 \sin 2\pi Nt.$$

The extra thrusts exerted by the springs are

$$\mu_1(z - a\theta - \delta_0 \sin 2\pi Nt), \quad \mu_2(z - \delta_0 \sin 2\pi Nt), \quad \mu_3(z + a\theta - \delta_0 \sin 2\pi Nt).$$

If the resistance to spring movement for each axle is kv, where v is the velocity of the axle-boxes in their guides, the total upward force on M_s due to friction is

$$3k \left(\frac{dz}{dt} - 2\pi N\delta_0 \cos 2\pi Nt \right).$$

Writing this in the form

$$4\pi n_a M_s \left(\frac{dz}{dt} - 2\pi N\delta_0 \cos 2\pi Nt \right),$$

the equation for the vertical movement of M_S then becomes

$$M_S\frac{d^2z}{dt^2}+4\pi n_d M_S\frac{dz}{dt}+(\mu_1+\mu_2+\mu_3)z$$
$$=(\mu_1+\mu_2+\mu_3)\delta_0\sin 2\pi Nt+8\pi^2 Nn_d M_S\delta_0\cos 2\pi Nt+(\mu_1-\mu_3)a\theta.$$

Writing $\qquad 4\pi^2 n_v{}^2=\dfrac{\mu_1+\mu_2+\mu_3}{M_S}\quad$ and $\quad\dfrac{\mu_1-\mu_3}{\mu_1+\mu_2+\mu_3}=\rho_1,$

the equation for z takes the form

$$\frac{d^2z}{dt^2}+4\pi n_d\frac{dz}{dt}+4\pi^2 n_v{}^2 z$$
$$=4\pi^2 n_v{}^2\delta_0\sin 2\pi Nt+8\pi^2 Nn_d\delta_0\cos 2\pi Nt+\rho_1 4\pi^2 n_v{}^2 a\theta.$$

The ratio ρ_1 in any practical case is certainly small, and provided that $a\theta$ does not become correspondingly large, the last term on the right-hand side of the equation will be of little account and under these circumstances the vertical motion of M_S will be unaffected by small angular displacements. How far it is legitimate to assume that these angular displacements are small must now be investigated.

For angular movements of M_S the equation of motion is

$$M_S k^2\frac{d^2\theta}{dt^2}=\mu_1 a\,(z-a\theta-\delta_0\sin 2\pi Nt)$$
$$-\mu_3 a\,(z+a\theta-\delta_0\sin 2\pi Nt)-4\pi n_d M_S\cdot\tfrac{2}{3}a^2\frac{d\theta}{dt}.$$

Writing $\qquad 4\pi^2 n_\theta{}^2=\dfrac{\mu_1+\mu_2}{M_S k^2}a^2,\quad \rho_2=\dfrac{\mu_1-\mu_3}{\mu_1+\mu_3}\quad$ and $\quad\dfrac{2}{3}\dfrac{a^2}{k^2}n_d=n_d{}',$

the equation for θ takes the form

$$\frac{d^2\theta}{dt^2}+4\pi n_d{}'\frac{d\theta}{dt}+4\pi^2 n_\theta{}^2\theta=4\pi^2 n_\theta{}^2\rho_2\left(\frac{z}{a}-\frac{\delta_0}{a}\sin 2\pi Nt\right).$$

Combining this with the equation for z previously stated, expressions for $a\theta$ and z in terms of δ_0 can be deduced, and it will be found that the semi-travel of $a\theta$ has the value

$$\rho_2 N^2 n_\theta{}^2\delta_0/\{[N^4-N^2(n_\theta{}^2+n_v{}^2+4n_d n_d{}')+n_\theta{}^2 n_v{}^2(1-\rho_1\rho_2)]^2$$
$$+[16N^3(n_d+n_d{}')-4N(n_d n_\theta{}^2+n_d{}'n_v{}^2)]^2\}^{\frac{1}{2}}.$$

If ρ_2 is small, as will always be the case in practice, $a\theta$ will also be small; it will develop peaks for frequencies which make

$$N^4-N^2(n_\theta{}^2+n_v{}^2+4n_d n_d{}')+n_\theta{}^2 n_v{}^2(1-\rho_1\rho_2)=0,$$

that is, when $N=n_\theta$ or n_v approximately, but even these peak values will be small compared with δ_0, owing to the small value of ρ_2 and the large values of n_d and $n_d{}'$ required to account for the heavy damping in the springs of a locomotive.

In the absence of damping, large values of $a\theta$ would certainly be developed when $N = n_v$ or n_θ, but, in the case of a locomotive, the heavy damping in the springs eliminates this possibility and the term $\rho_1 4\pi^2 n_v^2 a\theta$, which appears on the right-hand side of the equation for z, cannot have any appreciable effect on the vertical movement of the c.g. of M_S and on the total variations of bridge load produced thereby.

Under these circumstances the idealization adopted in Fig. 29, Chapter IX, is justified. Thus, if the locomotive has six axles, provided that the spring coefficients $\dfrac{\mu_1}{W_1}$, $\dfrac{\mu_2}{W_2}$, ... $\dfrac{\mu_6}{W_6}$ are not very unequal, it is legitimate to idealize the locomotive into a concentrated mass M_S resting on a single spring whose natural frequency is given by

$$n_s = \frac{1}{2\pi} \sqrt{\frac{\mu_1 + \mu_2 + \dots + \mu_6}{M_S} g},$$

where μ_1, μ_2, etc. are expressed in tons per foot of deflection, and M_S is expressed in tons.

From this formula, no difficulty should be encountered in obtaining the numerical value of n_s for any given locomotive and, although variations are to be expected, an average value for this characteristic was found by the Bridge Stress Committee to be 3.

Frictional resistance to spring movement

This is the most elusive locomotive characteristic, and direct experimental evidence upon which estimates can be based is very scanty. The resistance to spring movement arises mainly from friction between the laminations of the springs, but it is augmented appreciably by the friction between the axle-boxes and their guides. From analysis of deflection records obtained by the Bridge Stress Committee, it would appear that an average value for the total frictional force resisting spring movement is 10 tons, and direct experiment in which deflections were recorded for increasing and decreasing loads has provided confirmation that this figure is of the right order of magnitude. That a large amount of friction exists is indicated by the fact that a locomotive, standing on one track of a double-track bridge, shews no sign of spring movement due to the passage of a locomotive along the other track, unless the bridge oscillations are very active. By noting the amplitude and frequency of the track oscillations which will just break down friction, the value of F can be deduced. The determination of spring friction for a standing locomotive should present no experimental difficulty, but an estimate of F thus obtained cannot be trusted to give the value of this characteristic for a locomotive in motion, under steam pressure. The pressure between the axle-boxes and their guides is certainly affected by the forces in the connecting-rods and side rods due to steam pressure and inertia, and, for a two-cylinder locomotive running at a speed of 5 revs. per sec., a rough estimate suggests that the increase in F, due to inertia and steam pressure, may be of the order of ± 3 tons.

For a locomotive in motion, the value of F could be deduced by recording simultaneously the amplitude of the axle-box movements in their guides and the vertical oscillations of the track as the locomotive moves across a bridge.

The formula for determining F in this case is as follows:

$$F = \frac{\pi^3 M_s}{g} \{d_0{}^2 N^4 - e_0{}^2 (N^2 - n_s{}^2)^2\}^{\frac{1}{2}} \text{ tons},$$

where d_0 is the semi-amplitude of the track oscillations measured in feet, e_0 is the semi-amplitude of the axle-box movements measured in feet, M_s is the spring-borne mass of the locomotive measured in tons, n_s is the natural frequency of M_s on its springs, N is the speed of the driving-wheels in revs. per sec.

A proof of this formula will now be given.

If the vertical oscillation of the track is given by $d_0 \sin 2\pi N t$, and z is the depth of M_s below its mean position, the displacement, e, of the axle-box in its guides is given by

$$\frac{d^2 e}{dt^2} + 4\pi n_d \frac{de}{dt} + 4\pi^2 n_s{}^2 e = 4\pi^2 N^2 d_0 \sin 2\pi N t,$$

where spring friction is represented by $4\pi n_d M_s \dfrac{de}{dt}$.

Hence
$$e = \frac{N^2 d_0 \sin (2\pi N t - \phi)}{\{(N^2 - n_s{}^2)^2 + (2N n_d)^2\}^{\frac{1}{2}}}$$

and
$$e_0 = \frac{N^2 d_0}{\{(N^2 - n_s{}^2)^2 + (2N n_d)^2\}^{\frac{1}{2}}},$$

that is,
$$2N n_d e_0 = \{d_0{}^2 N^4 - e_0{}^2 (N^2 - n_s{}^2)^2\}^{\frac{1}{2}}.$$

The frictional force has a primary harmonic component of magnitude $\dfrac{4}{\pi} F$, and, in the analysis, this appears in the form

$$4\pi n_d M_s \times 2\pi N e_0.$$

Accordingly, $$2Nn_d e_0 = \frac{F}{\pi^3 M_s},$$

and, consequently,

$$F = \pi^3 M_s \{d_0{}^2 N^4 - e_0{}^2 (N^2 - n_s{}^2)^2\}^{\frac{1}{2}} \text{ kinetic units,}$$

that is, $$F = \frac{\pi^3 M_s}{g} \{d_0{}^2 N^4 - e_0{}^2 (N^2 - n_s{}^2)^2\}^{\frac{1}{2}} \text{ tons,}$$

where d_0 and e_0 are measured in feet and M_s is measured in tons.

A comprehensive experimental investigation along the line just indicated is perhaps the most important advance which could now be made towards achieving enhanced precision in the calculation of bridge oscillations and dynamic stresses. At present the most serious obstacle is the paucity of exact information about the friction in spring movement. For long- and short-span bridges a knowledge of F is not required, but for medium-span bridges, the extent of the oscillations, particularly those of a resonant character, is largely determined by the value of this particular locomotive characteristic. For long-span bridges, the calculation of synchronous oscillations demands an exact knowledge of n_b, the characteristic defining bridge damping. It is along these two directions, namely the evaluation of n_b and F, that further full-scale experimental research is urgently required. Out of the list of ten bridge and locomotive characteristics enumerated at the commencement of this Chapter, these are the only two about which an element of uncertainty and guesswork at present exists.

CHAPTER XIV

DYNAMIC BENDING-MOMENTS AND SHEARING-FORCES

When the state of oscillation in a bridge has been determined, the computation of the corresponding dynamic bending-moments and shearing-forces is a comparatively straightforward task. The assumption that the mode of vibration is a pure sine curve, induced solely by the primary component of the applied forces, is justifiable when deflections have to be computed, but it does not represent the curve of bending with sufficient exactness to permit of bending-moments and shearing-forces being deduced by the simple processes of double and triple differentiation of this pure sine curve. For such calculations it is necessary to compute the contributions made by the various distributed and concentrated inertia forces separately, these forces being classified as follows:

(1) The distributed load $m\dfrac{d^2y}{dt^2}$ per unit length, due to the inertia of the bridge.

(2) The distributed load $4\pi n_b m \dfrac{dy}{dt}$ per unit length, due to the damping resistance.

(3) The concentrated operative forces, that is, the reactions between the wheels and the rails.

DYNAMIC BENDING-MOMENTS

Consider in the first instance the bending-moments produced by these three separate classifications and suppose that the oscillation which gives rise to the inertia forces is defined by

$$y = \delta_0 \sin\frac{\pi x}{l} \sin 2\pi N t.$$

Due to the inertia of the bridge, the load per unit length is

$$4\pi^2 N^2 m \delta_0 \sin\frac{\pi x}{l} \sin 2\pi N t,$$

and the corresponding sagging bending-moment is

$$4N^2 M_G l \delta_0 \sin\frac{\pi x}{l} \sin 2\pi N t.$$

Due to the damping resistance, the load per unit length is

$$-8\pi^2 N n_b m \delta_0 \sin\frac{\pi x}{l}\cos 2\pi N t,$$

and the corresponding bending-moment is

$$-8 N n_b M_G l \delta_0 \sin\frac{\pi x}{l}\cos 2\pi N t.$$

Thus it appears that the bending-moment due to damping is only the small fraction $\dfrac{2n_b}{N}$ of that due to inertia and, furthermore, these two effects are not superposed but differ by $90°$ in phase.

LONG-SPAN BRIDGES

For bridges of long span, in which the locomotive springs remain locked, the operative force consists of the hammer-blow combined with the inertia force necessary to give M its vertical acceleration. When the locomotive is passing the centre of the bridge this latter force has the magnitude

$$4\pi^2 N^2 M \delta_0 \sin 2\pi N t$$

approximately and produces a central bending-moment of magnitude

$$\pi^2 N^2 M l \delta_0 \sin 2\pi N t.$$

The contribution made to the operative force by hammer-blow is

$$P \sin (2\pi N t + \phi),$$

where ϕ is the angle by which the oscillations lag behind the hammer-blow.

For the case of outstanding practical importance, namely that of resonance, when the frequency of the hammer-blow coincides with the fundamental frequency of the bridge at the instant the locomotive is passing the centre of the span, the hammer-blow and the oscillations differ in phase by $90°$ approximately, and this circumstance, taken in conjunction with the relatively small magnitude of the force, makes it legitimate to neglect altogether the slight contribution to the central bending-moment made directly by the hammer-blow. For a similar reason the contribution made by the damping resistance is negligible, since in the case of a long-span bridge it is numerically very small owing to the minuteness of n_b and, furthermore, it is $90°$ out of phase with the much larger bending-moments due to the inertia of the bridge and locomotive.

Neglecting these two small and "out-of-phase" contributions, the maximum central dynamic bending-moment for which provision must be made has the magnitude $(4M_G + \pi^2 M) l \delta_0 N^2$, where δ_0 is the dynamical increment of the central deflection at the instant the locomotive is passing the centre of the span and N is the fundamental frequency of the bridge at this instant, that is

$$N^2 = \frac{\pi^2}{4l^3} \frac{EI}{M_G + 2M}.$$

If M and M_G are measured in tons, and l and δ_0 measured in feet, this maximum central bending-moment is $\dfrac{(4M_G + \pi^2 M) l \delta_0 N^2}{g}$ tons feet, and the total load uniformly distributed which would produce this same maximum central bending-moment is $\left(M_G + \dfrac{\pi^2}{4} M \right) \delta_0 N^2$ tons, taking g to have the value 32 f.s. units.

If the bending-moment is deduced from the curve of deflection

$$y = \delta_0 \sin \frac{\pi x}{l} \sin 2\pi N t$$

by the formula $$M = - EI \frac{d^2 y}{dx^2},$$

the value thus obtained for the central bending-moment is $\dfrac{\pi^2}{l^2} EI \delta_0$ and, since for the case of resonance under consideration,

$$\frac{\pi^2}{l^2} EI = (4M_G + 8M) l N^2,$$

this value can be written $(4M_G + 8M) l \delta_0 N^2$,

which is appreciably less than the true value

$$(4M_G + \pi^2 M) l \delta_0 N^2,$$

and bears out the warning previously given that the curve of deflection is not a sufficiently accurate representation to permit of bending-moments and shearing-forces being deduced by the processes of double and triple differentiation.

NUMERICAL EXAMPLE

Consider the particular case of a long-span bridge studied in Chapter VIII, in which $l = 270$ feet, $M_G = 450$ tons, $n_0 = 3$, $n_b = 0.12$, $M = 100$ tons, $P = 0.6 N^2$ tons. For this case δ_0, the maximum central dynamic deflection, is found to have the value 0.0174 foot, when $N = 2.4961$, resonance being established at the instant the locomotive is passing the centre of the span.

According to the formula given above, the value of the live-load allowance, to account for the central dynamic bending-moment, is

$$(450 + 2.4674 \times 100) \times 0.0174 \times 2.4961^2 \text{ tons} = 75.5 \text{ tons}.$$

The contributions which were left out of account are:

(1) A live-load allowance $2Nn_b M_G \delta_0$ tons, to account for the central bending-moment due to damping.

(2) A live-load allowance $2P$ tons, to account for the central bending-moment due to hammer-blow.

These two bending-moments are anti-phased so that they yield a resultant which calls for a live-load allowance $(2Nn_b M_G \delta_0 - 2P)$ tons, and for the case under consideration, this has the value 5·7 tons. While small in magnitude this live load is not to be directly superposed on the load of 75·5 tons previously calculated. The bending-moments accounted for by these loads are 90° out of phase and the resultant live load is consequently

$$(75 \cdot 5^2 + 5 \cdot 7^2)^{\frac{1}{2}} \text{ tons} = 75 \cdot 7 \text{ tons,}$$

which for all practical purposes is indistinguishable from the approximate estimate of 75·5 tons.

SINGLE-TRACK LONG-SPAN BRIDGES

The foregoing calculation shews that the increment of live load which will account for the maximum central dynamic bending-moment due to a single locomotive is given by the formula

$$\left(M_G + \frac{\pi^2}{4} M \right) \delta_0 N^2 \text{ tons,}$$

where N is the critical frequency which gives the maximum state of oscillation and is determined by

$$N = n_0 \left\{ \frac{M_G}{M_G + 2M} \right\}^{\frac{1}{2}},$$

δ_0 is the semi-amplitude of the corresponding oscillation measured in feet, M is the total mass of the locomotive, M_G is the total mass of the bridge, both measured in tons, and n_0 is the fundamental unloaded frequency of the bridge.

For double-heading, with two similar locomotives spaced a distance d apart having their hammer-blows in step, the corresponding formula for the live-load allowance is

$$\left[M_G + \frac{\pi^2}{2} M \frac{l-d}{l} \cos \frac{\pi d}{2l} \right] \delta_0 N^2 \text{ tons,}$$

where
$$N = n_0 \left\{ \frac{M_G}{M_G + 4M \cos^2 \frac{\pi d}{2l}} \right\}^{\frac{1}{2}}$$

and δ_0 is the semi-amplitude of the maximum state of oscillation thus produced.

DOUBLE-TRACK LONG-SPAN BRIDGES

With two similar locomotives, one on each track, having their hammer-blows in step, the formula for the live-load allowance for dynamic bending is

$$\left(M_G + \frac{\pi^2}{2} M\right) \delta_0 N^2 \text{ tons,}$$

where
$$N = n_0 \left\{\frac{M_G}{M_G + 4M}\right\}^{\frac{1}{2}}.$$

For double-heading, with two pairs of similar locomotives, the longitudinal spacing in each pair being d, the live-load allowance for dynamic bending is

$$\left(M_G + \pi^2 M \frac{l-d}{l} \cos \frac{\pi d}{2l}\right) \delta_0 N^2 \text{ tons,}$$

where
$$N = n_0 \left\{\frac{M_G}{M_G + 8M \cos^2 \dfrac{\pi d}{2l}}\right\}^{\frac{1}{2}}.$$

This last case assumes the extremely improbable contingency of all four locomotives having their hammer-blows exactly in step and the two pairs crossing at the centre of the bridge at the same critical speed. This is such an unlikely coincidence, that full allowance for it need hardly be made, and probably half the increment of live load thus evaluated would suffice.

MEDIUM-SPAN BRIDGES

When the oscillations generated by a locomotive, in crossing a bridge, are sufficiently violent to stimulate spring movement, the bridge, if over 50 feet span, is classified as a medium-span bridge. For bridges of this class, the dynamic bending-moments due to inertia and damping resistance are the same as those computed for bridges of long span, but since for bridges of medium span the maximum state of oscillation can be regarded as set up by a stationary skidding locomotive, the central operative force (see Chapter XI, p. 129) can be viewed as consisting of two components, namely

$$2\pi^2 M_G (n_0^2 - N^2) \delta_0 \sin 2\pi N t$$

and
$$2\pi^2 M_G (2N n_b) \delta_0 \sin \left(2\pi N t + \frac{\pi}{2}\right),$$

where the oscillation at the centre of the span is defined by $\delta = \delta_0 \sin 2\pi N t$.

The corresponding central bending-moments, measured in tons feet, are

$$\frac{\pi^2}{64} M_G (n_0^2 - N^2) l \delta_0 \sin 2\pi N t \quad \text{and} \quad \frac{\pi^2}{64} M_G (2N n_b) \delta_0 \sin \left(2\pi N t + \frac{\pi}{2}\right),$$

where M_G is measured in tons, δ_0 and l are measured in feet and g has the value 32 in f.s. units.

For the inertia of the bridge and for the damping resistance, the contributions to the central dynamic bending-moments are respectively

$$\frac{M_G}{8} N^2 l \delta_0 \sin 2\pi N t \quad \text{and} \quad -\frac{M_G}{8} 2N n_b l \delta_0 \cos 2\pi N t \text{ tons feet.}$$

These four components combine to give a central bending-moment alternating between the limits

$$\pm \frac{M_G}{8} \left\{ \left[N^2 + \frac{\pi^2}{8} (n_0^2 - N^2) \right]^2 + \left[\left(\frac{\pi^2}{8} - 1 \right) 2N n_b \right]^2 \right\}^{\frac{1}{2}} l \delta_0 \text{ tons feet,}$$

and, since $\left(\dfrac{\pi^2}{8} - 1 \right) 2N n_b$ is, in all practical cases, insignificant compared with $N^2 + \dfrac{\pi^2}{8} (n_0^2 - N^2)$, the value of the central alternating bending-moment may be taken as

$$\frac{M_G}{8} \left[N^2 + \frac{\pi^2}{8} (n_0^2 - N^2) \right] l \delta_0 \text{ tons feet,}$$

and the increment of live load which will account for this central dynamic bending-moment is

$$M_G \left[N^2 + \frac{\pi^2}{8} (n_0^2 - N^2) \right] \delta_0 \text{ tons,}$$

where M_G is measured in tons and δ_0 is measured in feet.

NUMERICAL EXAMPLE

For a particular case studied in Chapter X in which $l = 150$ feet, $M_G = 300$ tons, $n_0 = 6$, $n_b = 0.36$, and the total mass of the locomotive is 80 tons, it was found that when the frictional resistance in the spring movement was 8 tons, the maximum central dynamic deflection was 0.136 inch, and this occurred for a frequency $N = 6.11$.

The increment of live load which will account for the central dynamic bending-moment, as given by the formula above, is

$$300 \left[6.11^2 - \frac{\pi^2}{8} (6.11^2 - 6^2) \right] \frac{0.136}{12} \text{ tons} = 121.3 \text{ tons.}$$

The quadrature component which has been neglected, namely

$$M_G \left(\frac{\pi^2}{8} - 1 \right) 2N n_b \delta_0 \text{ tons,}$$

has the value 3·5 tons. The inclusion of this would only augment the live-load allowance to 121·35 tons, an increase which is utterly insignificant. Accordingly the formula

$$M_G\left[N^2-\frac{\pi^2}{8}(n_0{}^2-N^2)\right]\delta_0$$

may be accepted as giving the live-load allowance for dynamic bending with an accuracy which is sufficient for all practical purposes and the position may be summarized as follows.

SINGLE-TRACK MEDIUM-SPAN BRIDGES

Case (1)

The increment of live load which will account for the maximum central dynamic bending-moment due to a single locomotive is given by the formula

$$M_G\left[N_2{}^2+\frac{\pi^2}{8}(n_0{}^2-N_2{}^2)\right]\delta_0\ \text{tons},$$

where N_2 is the critical frequency which gives the maximum state of oscillation, and is determined by the equation

$$N_2{}^4-N_2{}^2n_1{}^2\left(1+\frac{n_s{}^2}{n_2{}^2}\right)+n_1{}^2n_s{}^2=0,$$

or, alternatively, if the value of N_2 thus obtained is beyond the range of practical speed, the largest practical value should be assigned, say $N_2=6$.

In the formula δ_0 is the semi-amplitude of the corresponding oscillation measured in feet, and M_G is the total mass of the bridge measured in tons.

Case (2)

For double-heading, with two similar locomotives spaced a distance d feet apart having their hammer-blows in step, each operative force has the components

$$\pi^2M_G(n_0{}^2-N_2{}^2)\ \frac{\delta_0}{\cos\dfrac{\pi d}{2l}}\sin 2\pi Nt$$

and

$$\pi^2M_G(2N_2n_b)\ \frac{\delta_0}{\cos\dfrac{\pi d}{2l}}\sin\left(2\pi Nt+\frac{\pi}{2}\right),$$

and, neglecting this second component as giving a negligible contribution, the formula for maximum live-load allowance is

$$M_G\left[N_2{}^2+\frac{\pi^2}{8}\frac{l-d}{l}\frac{1}{\cos\dfrac{\pi d}{2l}}(n_0{}^2-N_2{}^2)\right]\delta_0\ \text{tons}.$$

DOUBLE-TRACK MEDIUM-SPAN BRIDGES

Case (3)

With two similar locomotives, one on each track, having their hammer-blows in step, the formula for the live-load allowance for dynamic bending-moment is

$$M_G\left[N_2{}^2+\frac{\pi^2}{8}(n_0{}^2-N_2{}^2)\right]\delta_0 \text{ tons.}$$

Case (4)

For double-heading, with two pairs of similar locomotives, the locomotives in each pair being spaced a distance d feet apart, the live-load allowance for dynamic bending-moment is given by the formula

$$M_G\left[N_2{}^2+\frac{\pi^2}{8}\frac{l-d}{l}\frac{1}{\cos\dfrac{\pi d}{2l}}(n_0{}^2-N_2{}^2)\right]\delta_0 \text{ tons,}$$

but this last case, as in the corresponding case of a long-span bridge, pre-supposes such an improbable coincidence of circumstances that the full live-load allowance given above need hardly be enforced. The values of δ_0 will naturally vary in accordance with the number of locomotives in action and the frequency which corresponds to the maximum states of oscillation, as given by the equation

$$N_2{}^4-N_2{}^2n_1{}^2\left(1+\frac{n_s{}^2}{n_2{}^2}\right)+n_1{}^2n_s{}^2=0,$$

will also be different for the four cases which have been considered.

Thus for Case (1)

$$n_1{}^2=n_0{}^2\frac{M_G}{M_G+2M_U},\quad n_2{}^2=n_0{}^2\frac{M_G}{M_G+2(M_U+M_s)}.$$

Case (2)

$$n_1{}^2=n_0{}^2\frac{M_G}{M_G+4M_U\cos^2\dfrac{\pi d}{2l}},\quad n_2{}^2=n_0{}^2\frac{M_G}{M_G+4(M_U+M_s)\cos^2\dfrac{\pi d}{2l}}.$$

Case (3)

$$n_1{}^2=n_0{}^2\frac{M_G}{M_G+4M_U},\quad n_2{}^2=n_0{}^2\frac{M_G}{M_G+4(M_U+M_s)}.$$

Case (4)

$$n_1{}^2=n_0{}^2\frac{M_G}{M_G+8M_U\cos^2\dfrac{\pi d}{2l}},\quad n_2{}^2=n_0{}^2\frac{M_G}{M_G+8(M_U+M_s)\cos^2\dfrac{\pi d}{2l}}.$$

SHORT-SPAN BRIDGES

This classification includes all bridges of less than 50 feet span. In Chapter XII it was shewn that owing to the large amount of damping and the high natural frequency inherent in bridges of this class, the operating forces approximate in magnitude to the hammer-blows, while in phase they are in advance of the oscillation by an angle ϕ, where

$$\tan \phi = \frac{2Nn_b}{n_0{}^2 - N^2}.$$

What numerical value should be assigned to n_b for a short-span bridge is largely a matter of guesswork, but fortunately for calculations of dynamic deflections in short-span bridges this information is not required, and an excellent approximation to the live-load allowance for dynamic bending-moment can be obtained without having to take this particular bridge characteristic into account.

For a locomotive, standing on the bridge and skidding its wheels at a constant speed N, let $R_1 \sin(2\pi Nt + \phi)$, $R_2 \sin(2\pi Nt + \phi)$, etc. be the operative forces expressed in tons, acting at distances a_1, a_2, etc. from one end of the bridge.

The dynamic deflection at the centre of the bridge will be given by

$$\delta = \frac{16}{\pi^2 M_G} \frac{\left(R_1 \sin \frac{\pi a_1}{l} + R_2 \sin \frac{\pi a_2}{l} + \ldots\right) \sin 2\pi Nt}{\{(n_0{}^2 - N^2)^2 + (2Nn_b)^2\}^{\frac{1}{2}}}.$$

The central bending-moment due to the operative forces alternates between the limits

$$\pm \frac{2l}{\pi^2}\left(R_1 \sin \frac{\pi a_1}{l} + R_2 \sin \frac{\pi a_2}{l} + \ldots\right) \text{ tons feet approximately,}$$

and, if W_P is the uniformly distributed load which gives this same central bending-moment,

$$W_P = \frac{16}{\pi^2}\left(R_1 \sin \frac{\pi a_1}{l} + R_2 \sin \frac{\pi a_2}{l} + \ldots\right) \text{ tons.}$$

Hence δ_0, the semi-amplitude of the central dynamic deflection, is given by

$$\delta_0 = \frac{W_P}{M_G\{(n_0{}^2 - N^2)^2 + (2Nn_b)^2\}^{\frac{1}{2}}} \text{ feet,}$$

W_P and M_G being expressed in tons.

The central bending-moments due to inertia and damping resistance in the bridge call for live-load allowances of

$$M_G N^2 \delta_0 \sin 2\pi Nt \quad \text{and} \quad -M_G 2Nn_b \delta_0 \cos 2\pi Nt \text{ tons}$$

respectively. Combining these various components, the total live-load allowance for the central dynamic bending-moment is

$$\{(W_P\cos\phi+M_G N^2\delta_0)^2+(W_P\sin\phi-M_G 2Nn_b\delta_0)^2\}^{\frac{1}{2}} \text{ tons.}$$

Since

$$\sin\phi=\frac{2Nn_b}{\{(n_0^2-N^2)^2+(2Nn_b)^2\}^{\frac{1}{2}}}, \quad \cos\phi=\frac{n_0^2-N^2}{\{(n_0^2-N^2)^2+(2Nn_b)^2\}^{\frac{1}{2}}}$$

and

$$\delta_0=\frac{W_P}{M_G\{(n_0^2-N^2)^2+(2Nn_b)^2\}^{\frac{1}{2}}},$$

it follows that

$$W_P\sin\phi-M_G 2Nn_b\delta_0=0$$

and

$$W_P\cos\phi+M_G N^2\delta_0=\frac{n_0^2}{\{(n_0^2-N^2)^2+(2Nn_b)^2\}^{\frac{1}{2}}}W_P.$$

But $\dfrac{n_0^2}{\{(n_0^2-N^2)^2+(2Nn_b)^2\}^{\frac{1}{2}}}$ is the dynamic magnification for the oscillations and, for short spans in which n_0 is large compared with N, this magnification approximates to unity. Consequently, the live-load allowance for the central dynamic bending-moment reduces to the pleasingly simple result W_P tons. Furthermore, since the operative forces have been found to have approximately the same magnitude as the hammer-blows, W_P may be regarded as the equivalent uniformly distributed load for the hammer-blows treated as statical forces.

Numerical example

Bridge data. Span 30 feet, $M_G=40$ tons, $n_0^2=200$, $n_b=6\cdot0$. The hammer-blows at 6 revs. per sec. and the axle spacings are as shewn in Fig. 53, Chapter XII.

The maximum central bending-moment for the hammer-blows treated as statical forces amounts to 144 tons feet approximately, and the equivalent uniformly distributed load W_P has the value 38·4 tons. The central dynamic deflection for this position of the locomotive was found to have the value 0·00455 foot, and consequently $M_G N^2\delta_0=6\cdot55$ and $M_G 2Nn_b\delta_0=13\cdot10$. From these results it follows that, for the dynamic bending-moment due to the operative forces, the live-load allowance is $38\cdot4\sin(2\pi Nt+\phi)$ tons, the allowance for the bending-moment due to inertia in the bridge is $6\cdot55\sin 2\pi Nt$ tons and the allowance for the bending-moment due to the damping resistance is $-13\cdot10\cos 2\pi Nt$; also

$$\sin\phi=\frac{2Nn_b}{\{(n_0^2-N^2)^2+(2Nn_b)^2\}^{\frac{1}{2}}}=0\cdot4, \quad \cos\phi=\frac{n_0^2-N^2}{\{(n_0^2-N^2)^2+(2Nn_b)^2\}^{\frac{1}{2}}}=0\cdot91.$$

Hence the two components of the live-load allowance are

$$38\cdot4 \times 0\cdot91 + \ 6\cdot55 = 41\cdot5 \ \text{tons}$$

and
$$38\cdot4 \times 0\cdot40 - 13\cdot10 = \ 2\cdot3 \ \text{tons}.$$

The resultant $(41\cdot5^2 + 2\cdot3^2)^{\frac{1}{2}}$ has the value $41\cdot6$ tons. Thus it appears that the approximate estimate, $W_P = 38\cdot4$ tons, slightly under-states the allowance, but, for all practical purposes, the discrepancy is negligible. Such error as exists is due to the fact that the dynamic magnification even for a 30 feet span is somewhat in excess of unity, and with the particular characteristics assumed for this case the magnification is approximately $\frac{10}{9}$.

The conclusion reached may be summarized as follows.

For a short-span bridge, the live-load allowance for dynamic bending-moments can be taken as a uniformly distributed load W_P, where W_P is the equivalent uniformly distributed load for the hammer-blows of the various axles of a locomotive, treating these blows as statical forces and calculating them for the highest permissible speed.

DYNAMIC SHEARING-FORCES

LONG-SPAN BRIDGES

If, at the moment under consideration, the dynamic deflection of the bridge is defined by
$$y = \delta_0 \sin \frac{\pi x}{l} \sin 2\pi Nt,$$

the load distribution due to inertia is

$$4\pi^2 N^2 m \delta_0 \sin \frac{\pi x}{l} \sin 2\pi Nt \ \text{per unit length},$$

and the corresponding distribution of shearing-force is

$$4\pi M_G N^2 \delta_0 \cos \frac{\pi x}{l} \sin 2\pi Nt.$$

For the damping resistance, the distribution of shearing-force is given by

$$-4\pi M_G 2N n_b \delta_0 \cos \frac{\pi x}{l} \cos 2\pi Nt,$$

and, in the case of long-span bridges, this is reduced to insignificance owing to the small numerical value of n_b. The value of δ_0 will be greatest when the locomotive is near the centre of the bridge moving at a speed which makes $N^2 = n_0^2 \dfrac{M_G}{M_G + 2M}$; but, even when it has reached the end of the span, δ_0 in the case of a long-span bridge may still be a considerable percentage of the maximum and, to be on the safe side, it will be assumed that while the locomotive is moving from the centre to the end of the span the value of δ_0 remains constant.

The distribution of shearing-force due to the inertia of the bridge has the formation shewn in Fig. 60. The shearing-force is zero at the centre of the span and attains its maximum $4\pi M_G N^2 \delta_0$ at the ends. To this must be added the shear due to the operative forces. Assuming that the springs are locked by friction, the operative force, when the locomotive is at a distance a from the further end of the span, consists of a component

$$4\pi^2 N^2 M \delta_0 \sin \frac{\pi a}{l} \sin 2\pi N t$$

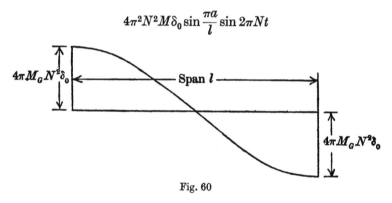

Fig. 60

in phase with the oscillations, combined with the hammer-blow which, for the extreme synchronous case under consideration, is approximately 90° out of phase with the oscillations. This circumstance, combined with the fact that the hammer-blow corresponding to this case of synchronism is comparatively small in magnitude, justifies its omission in calculating dynamic shear. For the inertia force due to the mass of the locomotive, the maximum shearing-force at any section occurs when the locomotive is passing that section and its value is $4\pi^2 N^2 M \frac{a}{l} \sin \frac{\pi a}{l} \delta_0$, where a is the distance of the section from the further end of the span. Assuming δ_0 is constant for the second half of the run, the maximum shearing-force diagram for the shear due to the inertia of the locomotive has the double peaked formation indicated by Fig. 61.

Combining the distributions illustrated by Figs. 60 and 61, the maximum shearing-force at a section distant a from the further end of the span is given by the expression

$$\frac{\pi}{8}\left[-M_G \cos \frac{\pi a}{l} + M \frac{\pi a}{l} \sin \frac{\pi a}{l} \right] N^2 \delta_0 \text{ tons,}$$

where M_G and M are measured in tons, δ_0 is the maximum central dynamical deflection measured in feet, N is the frequency of the hammer-blows corresponding to the maximum state of oscillation and g is taken to have the value 32 in f.s. units.

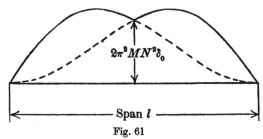

Fig. 61

It is not possible to make provision for this maximum shearing-force by any system of equivalent loads uniformly distributed or concentrated; but, for any particular case, the maximum dynamical shearing-force which has to be allowed for at any section can be readily calculated from the formula stated above, and this process will now be illustrated by the following numerical example.

NUMERICAL EXAMPLE

Consider the particular case of a long-span bridge studied in Chapter VIII, in which $l = 270$ feet, $M_G = 450$ tons, $M = 100$ tons, $P = 0 \cdot 6 N^2$ tons, and δ_0 has its maximum value of $0 \cdot 0174$ foot when $N = 2 \cdot 4961$.

The formula for the maximum dynamic shearing-force at a section distant a from the further end of the span, namely

$$\frac{\pi}{8}\left[-M_G \cos\frac{\pi a}{l} + M\frac{\pi a}{l}\sin\frac{\pi a}{l} \right] N^2 \delta_0 \text{ tons,}$$

gives in this case

$$-19 \cdot 05 \cos\frac{\pi a}{l} + 13 \cdot 30\frac{a}{l}\sin\frac{\pi a}{l} \text{ tons,}$$

and the maximum dynamic shearing-force diagram has the form shewn in Fig. 62.

Fig. 62 gives the maximum shearing-force for the case when the oscillations are most violent, and in deciding allowances for shear the only other case which demands consideration is that in which the locomotive is crossing the bridge at the highest permissible speed (say $N = 6$).

For the case just considered, when $N = 6$, the hammer-blow has the high value of $21 \cdot 6$ tons, but the maximum central dynamical deflection is only $0 \cdot 0025$ foot approximately. The inertia force of the locomotive when it is passing the centre of the bridge is about 11 tons, and since at this high speed hammer-blow and the oscillations are anti-phased the operative force is reduced to about $10\frac{1}{2}$ tons, which gives a smaller shear stress at the centre than that obtained for the case $N = 2\cdot4961$.

At the instant the locomotive comes on to the bridge, the end shearing-force is $21\cdot6$ tons, which slightly exceeds the value $19\cdot05$ obtained for maximum end shear for the case $N = 2\cdot4961$, but the discrepancy is hardly

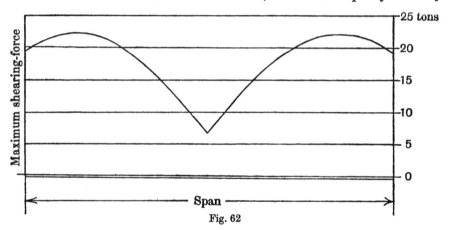

Fig. 62

worth taking into account and, for practical purposes, the allowance for shear required to deal with the case of maximum oscillation may be regarded as sufficient to cover all other possible cases.

Summary

SINGLE-TRACK LONG-SPAN BRIDGES

For a single locomotive the maximum dynamic shearing-force at a section distant a from the further end of the span has the value

$$\frac{\pi}{8}\left[-M_G\cos\frac{\pi a}{l} + M\frac{\pi a}{l}\sin\frac{\pi a}{l}\right]N^2\delta_0 \text{ tons.}$$

For double-heading with two similar locomotives spaced a distance d apart the corresponding formula is

$$\frac{\pi}{8}\left[-M_G\cos\frac{\pi a}{l} + M\frac{\pi a}{l}\sin\frac{\pi a}{l} + M\frac{\pi(a-d)}{l}\sin\frac{\pi(a-d)}{l}\right]N^2\delta_0 \text{ tons,}$$

where M and M_G are measured in tons, δ_0 is the maximum central dynamic deflection (measured in feet) produced by the single locomotive in the first case and by the two locomotives in the second, N is the frequency corresponding to the maximum state of oscillation and g is taken to have the value 32 in f.s. units.

DOUBLE-TRACK LONG-SPAN BRIDGES

These cases can be dealt with by writing $2M$ in place of M in the formulae for single-track bridges; thus for two similar locomotives, one on each track, the maximum dynamic shearing-force at a section distant a from the further end of the span has the value

$$\frac{\pi}{8}\left[-M_G\cos\frac{\pi a}{l}+2M\frac{\pi a}{l}\sin\frac{\pi a}{l}\right]N^2\delta_0\text{ tons.}$$

For double-heading on each track with four similar locomotives in pairs, the longitudinal spacing for each pair being defined by d, the corresponding formula for the maximum dynamic shearing-force is

$$\frac{\pi}{8}\left[-M_G\cos\frac{\pi a}{l}+2M\frac{\pi a}{l}\sin\frac{\pi a}{l}+2M\frac{\pi(a-d)}{l}\sin\frac{\pi(a-d)}{l}\right]N^2\delta_0\text{ tons,}$$

where M and M_G are measured in tons, δ_0 is the maximum central dynamic deflection (measured in feet) produced by the two locomotives in the first case and by the four locomotives in the second, N is the frequency corresponding to the maximum state of oscillation and g is taken to have the value 32 in f.s. units.

In the last case the greatest value of δ_0 would be obtained by considering the hammer-blow of all four locomotives to be in step, but the probability of this is so remote that full allowance for this unlikely contingency need hardly be provided.

MEDIUM-SPAN BRIDGES

Allowance for dynamic shear

For bridges of this class the maximum state of oscillation is developed when the locomotive is at the centre of the span and its hammer-blows have the critical frequency N_2, where N_2^2 is the larger of the two roots of the equation

$$N_2^4-N_2^2n_1^2\left(1+\frac{n_s^2}{n_2^2}\right)+n_1^2n_s^2=0.$$

If the critical value thus determined is beyond the range of practical speed, the largest attainable value of N (say 6) must be employed for computing the maximum oscillations and, in either case, this frequency will be denoted by N_2.

For the inertia of the girder the shearing-force at a section distant a from the further end of the span, when the oscillation is defined by

$$y=\delta_0\sin\frac{\pi x}{l}\sin 2\pi N_2t,$$

is

$$-4\pi M_G\cos\frac{\pi a}{l}\times\frac{N_2^2\delta_0}{g}\sin 2\pi N_2t\text{ tons,}$$

and the corresponding shearing-force due to damping resistance has the value

$$4\pi M_G \cos\frac{\pi a}{l} \cdot \frac{2N_2 n_b \delta_0}{g} \cos 2\pi N_2 t \text{ tons.}$$

If R is the central operative force measured in tons,

$$\delta_0 = \frac{1}{2\pi^2 M_G} \cdot \frac{Rg}{\{(n_0^2 - N_2^2)^2 + (2N_2 n_b)^2\}^{\frac{1}{2}}} \text{ feet,}$$

and the oscillations lag behind the operative force by an angle ϕ, where

$$\tan\phi = \frac{2N_2 n_b}{n_0^2 - N_2^2}.$$

Consequently, the component of the operative force which is in phase with the oscillations is

$$\frac{2\pi^2 M_G}{g}(n_0^2 - N_2^2)\,\delta_0 \sin 2\pi N_2 t \text{ tons,}$$

and the component in quadrature is

$$\frac{2\pi^2 M_G}{g}(2N_2 n_b)\,\delta_0 \cos 2\pi N_2 t \text{ tons.}$$

The corresponding shearing-forces at any section are respectively

$$\pm\frac{\pi^2 M_G}{g}(n_0^2 - N_2^2)\,\delta_0 \sin 2\pi N_2 t \quad \text{and} \quad \pm\frac{\pi^2 M_G}{g}(2N_2 n_b)\,\delta_0 \cos 2\pi N_2 t \text{ tons.}$$

Combining these with the shearing-forces due to the inertia of the bridge and damping resistance, the maximum shearing-force at the section a from the further end of the span is given by the expression

$$\frac{\pi M_G \delta_0}{g}\left[-4N_2^2 \cos\frac{\pi a}{l} + \pi(n_0^2 - N_2^2)\right]\sin 2\pi N_2 t$$

$$+ \frac{\pi M_G \delta_0}{g} \cdot 2N_2 n_b \left[4\cos\frac{\pi a}{l} + \pi\right]\cos 2\pi N_2 t.$$

If this is written in the form

$$S_1 \sin 2\pi N_2 t + S_2 \cos 2\pi N_2 t,$$

the intensity of the maximum dynamic shearing-force at the section distant a from the further end of the span has the value $(S_1^2 + S_2^2)^{\frac{1}{2}}$ tons, where M_G is expressed in tons and δ_0 in feet units.

For a complete examination of the maximum dynamical shearing-force, various positions of the locomotive, other than the central position, should be studied, and for each of these positions the maximum obtainable value of δ_0 should be determined. This detailed investigation is, however, hardly necessary. The case considered, in which the locomotive is skidding its wheels at the centre of the bridge, with the critical speed, or alternatively, the highest possible speed, will certainly give the maximum shearing-force at the ends of the span and, for intermediate sections, this method of determining the maximum dynamic shear only errs in under-stating the value in the immediate neighbourhood of the centre of the span.

This point is brought out by the following numerical example.

NUMERICAL EXAMPLE

Consider the particular example of a medium-span bridge studied in Chapter X in which $l = 150$ feet, $M_G = 300$ tons, $n_0 = 6$, $n_b = 0.36$, $M_S = 80$ tons, $M_U = 20$ tons, $F = 8$ tons, $P = 0.6N^2$ tons, $n_s = 3$. For the locomotive skidding its wheels at the centre of the bridge, the central dynamic deflection was found to attain its maximum value 0.136 inch when $N = 6.11$.

For this case, the dynamic shearing-force at the section distant a from the further end of the span is given by the expression

$$\left[-49.84 \cos \frac{\pi a}{l} - 1.40 \right] \sin 2\pi N_2 t + \left[5.87 \cos \frac{\pi a}{l} + 4.62 \right] \cos 2\pi N_2 t \text{ tons.}$$

The contribution $-49.84 \cos \dfrac{\pi a}{l} \sin 2\pi N_2 t$ is due to the inertia of the girder and it will be seen that, except when a is nearly $\dfrac{l}{2}$, it far outweighs the other contributions.

Writing the expression above in the form $S_1 \sin 2\pi N_2 t + S_2 \cos 2\pi N_2 t$, the values of S_1 and S_2 for various values of $\dfrac{a}{l}$ are tabulated below, and Fig. 63 shews how the value of $(S_1^2 + S_2^2)^{\frac{1}{2}}$, the maximum dynamic shearing-force, varies along the span.

$\dfrac{a}{l}$	1	$\dfrac{15}{16}$	$\dfrac{7}{8}$	$\dfrac{13}{16}$	$\dfrac{3}{4}$	$\dfrac{11}{16}$	$\dfrac{5}{8}$	$\dfrac{9}{16}$	$\dfrac{1}{2}$
S_1	48.44	47.48	44.65	40.04	33.84	26.29	17.67	8.32	-1.40
S_2	-1.25	-1.14	-0.80	-0.26	$+0.47$	$+1.36$	$+2.38$	$+3.47$	$+4.62$
$(S_1^2 + S_2^2)^{\frac{1}{2}}$	48.5	47.5	44.7	40.1	33.9	26.5	17.8	9.0	4.8

In this case, owing to the fact that N_2 and n_0 are nearly equal in magnitude, the intensity of the central operative force is reduced to 9.6 tons, and considerably greater values of the force can be obtained at lower speeds. This reduction in the intensity of the operative force is brought about by the vertical movements of M_S on its springs, and the operative force approximates to its maximum intensity when oscillations in the bridge are only just sufficiently vigorous to stimulate spring movement.

If the frequency corresponding to this state of affairs is denoted by N_F, it was shewn in Chapter X, page 121 that the equation for finding N_F is

$$N_F{}^4 = \frac{F}{2P_1} \cdot \frac{M_G}{M_S} n_0{}^2 \left\{ \left(1 - \frac{N_F{}^2}{n_2{}^2}\right)^2 + \left(\frac{2N_F n_b}{n_0{}^2}\right)^2 \right\}^{\frac{1}{2}},$$

where F, M_G and M_S are measured in tons, and P_1 tons is the hammer-blow at 1 rev. per sec.

For the numerical case just considered, $N_F = 4$ approximately, and the corresponding hammer-blow is 9·6 tons. The contribution to the operative force made by the inertia of the locomotive (springs locked) is $\dfrac{4\pi^2 M \delta_0 N^2}{g}$ tons

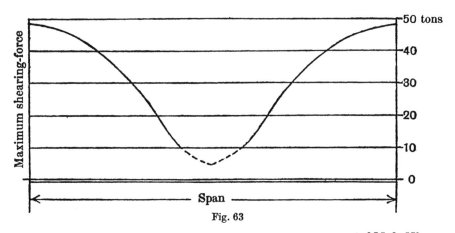

Fig. 63

and, since the breakdown of spring friction occurs when $\dfrac{4\pi^2 M_S \delta_0 N^2}{g} = F$,

this force has the magnitude $\dfrac{M}{M_S} F$ tons, which for this particular case gives the value 10 tons. These two components of 9·6 and 10 tons are out of phase by an angle ϕ, where $\tan\phi = \dfrac{2N_F n_b}{n_0{}^2 - N_F{}^2} = 0\cdot144$, but this angle is so small that their vector resultant is practically their arithmetic sum, namely 19·6 tons, which is rather more than twice the value of the central operative force at the critical frequency 6·11.

Consequently, although the value $(S_1{}^2 + S_2{}^2)^{\frac{1}{2}}$ previously obtained for the case of the locomotive skidding its wheels at the centre of the bridge gives the maximum dynamic shear at the ends and for the greater part of the span, the value in the region near the centre should be taken as $\dfrac{1}{2}\left(P_1 N_F{}^2 + \dfrac{M}{M_S} F\right)$, where N_F is the frequency at which friction in the spring of the central skidding locomotive will just break down, and $P_1 N_F{}^2$ tons is the corresponding hammer-blow. This point is indicated in Fig. 63 by dotting the part of the curve which falls below the level 9·6 tons.

Summary

ALLOWANCE FOR DYNAMIC SHEAR IN MEDIUM-SPAN BRIDGES

SINGLE-TRACK BRIDGES

For a single locomotive the dynamic shear to be allowed for at a section distant a from the further end of the span is $(S_1{}^2 + S_2{}^2)^{\frac{1}{2}}$ tons, where

$$S_1 = \frac{\pi M_G \delta_0}{32}\left[-4N_2{}^2 \cos\frac{\pi a}{l} + \pi\left(n_0{}^2 - N_2{}^2\right) \right],$$

$$S_2 = \frac{\pi M_G \delta_0}{32} \cdot 2N_2 n_b\left(4\cos\frac{\pi a}{l} + \pi\right),$$

where M_G is measured in tons and δ_0, measured in feet, is the maximum central dynamical deflection which can be produced by the locomotive, the corresponding speed of revolutions of its driving-wheels being N_2 revs. per sec.

For the region at the centre of the span the values thus obtained must not be allowed to fall below the value $\frac{1}{2}\left(P_1 N_F{}^2 + \frac{M}{M_s}F\right)$ tons, where N_F is the frequency at which spring friction begins to break down, $P_1 N_F{}^2$ tons is the corresponding hammer-blow, M is the total mass of the locomotive in tons, M_s is the mass of its spring-borne part in tons, and F is the force resisting spring motion measured in tons.

For double-heading, with two similar locomotives spaced a distance d apart, the corresponding values of S_1 and S_2 are

$$S_1 = \frac{\pi M_G \delta_0}{32}\left[-4N_2{}^2\cos\frac{\pi a}{l} + \frac{\pi}{\cos\dfrac{\pi d}{2l}}\left(n_0{}^2 - N_2{}^2\right) \right],$$

$$S_2 = \frac{\pi M_G \delta_0}{32} \cdot 2N_2 n_b\left[4\cos\frac{\pi a}{l} + \frac{\pi}{\cos\dfrac{\pi d}{2l}} \right],$$

the lower limit in this case being $P_1 N_F{}^2 + \dfrac{M}{M_s}F$ tons.

DOUBLE-TRACK BRIDGES

For two similar locomotives one on each track passing at the centre with their hammer-blows in step, the dynamic shear to be allowed for at a section distant a from the further end of the span is $(S_1{}^2 + S_2{}^2)^{\frac{1}{2}}$ tons, where

$$S_1 = \frac{\pi M_G \delta_0}{32}\left[-4N_2{}^2 \cos\frac{\pi a}{l} + \pi (n_0{}^2 - N_2{}^2) \right],$$

$$S_2 = \frac{\pi M_G \delta_0}{32}. 2N_2 n_b \left(4\cos\frac{\pi a}{l} + \pi \right),$$

the lower limit being $P_1 N_F{}^2 + \dfrac{M}{M_s} F$.

For double-heading on each track, the four similar locomotives having their hammer-blows in step and the spacing for each pair being d,

$$S_1 = \frac{\pi M_G \delta_0}{32}\left[-4N_2{}^2 \cos\frac{\pi a}{l} + \frac{\pi}{\cos\dfrac{\pi d}{2l}} (n_0{}^2 - N_2{}^2) \right],$$

$$S_2 = \frac{\pi M_G \delta_0}{32}\left[4\cos\frac{\pi a}{l} + \frac{\pi}{\cos\dfrac{\pi d}{2l}} \right],$$

the lower limit in this case being $2P_1 N_F{}^2 + \dfrac{2M}{M_s} F$.

The formulae for S_1 and S_2 which deal with double-heading are not, strictly speaking, applicable to sections where $a < \dfrac{l+d}{2}$. In these cases the operative forces causing the maximum state of oscillation are equal and equally spaced on either side of the centre, so that, for a length d in the middle of the span, the operative forces make no direct contribution to shear. For slightly unsymmetrical positions these operative forces will develop shear in the middle length d and the contributions will differ but little from that obtained in the formulae for S_1 and S_2. The discrepancy is too small to be of any practical importance, particularly in view of the fact that for this central region the dominating condition is usually the statement that provision for dynamic shear in the two cases must not be less than $P_1 N_F{}^2 + \dfrac{M}{M_s} F$ and $2P_1 N_F{}^2 + \dfrac{2M}{M_s} F$ respectively.

It will be seen that the expressions for S_1 and S_2 for the single- and double-track bridges are identical in form, but the values of N_2 and δ_0 will be different. The value of N_2 is either the higher critical frequency given by the equation

$$N^4 - N^2 n_1{}^2 \left(1 + \frac{n_s{}^2}{n_2{}^2}\right) + n_1{}^2 n_s{}^2 = 0,$$

or, if this gives an impossibly large value, N_2 must be taken to be that corresponding to the highest permissible engine speed. The values of $n_1{}^2$ and $n_2{}^2$ which appear in equation for the critical frequency will vary in accordance with the type of loading, and these values have already been stated on p. 176.

The methods for determining the value of δ_0 corresponding to any value of N_2 have been explained in Chapters X and XI.

SHORT-SPAN BRIDGES

ALLOWANCE FOR DYNAMIC SHEAR

For bridges of this class it has been shewn that the operative forces approximate closely in magnitude to the hammer-blows and are almost non-dynamic in their effect. Under these circumstances the dynamic shearing-force at any section is prescribed almost entirely by the operative forces and its determination becomes a statical rather than a dynamical computation.

Accordingly, in so far as the operative force can be regarded as having the same intensities as the hammer-blows, the maximum shearing-force diagram can be computed with sufficient accuracy by the simple process of adding the hammer-blows appropriate to the highest permissible engine speed to the corresponding statical axle loads.

EFFECT OF A FOLLOWING TRAIN

In this and in the preceding Chapters, the dynamic effects of locomotives have been considered, without any reference to the train to which they may be attached. For prescribing dynamic allowances this is all that is necessary, since theory and experiment combine to shew that the effect of a following train is to reduce somewhat the extent of the oscillations, and the tender of a locomotive has a similar effect.

Even in the case of long-span bridges, the springs of the tender, passenger coaches or goods-wagons constituting a train will certainly be set in motion when the oscillations are resonant in character and the energy dissipated in overcoming spring friction increases the apparent damping of the bridge to an appreciable extent. In Chapter IX it was shewn that even a massive locomotive, provided its springs were stimulated into action, had practically no influence on the natural frequency of the bridge, the loaded and unloaded fundamental frequencies, under these circumstances, being almost identical. Accordingly, it may be inferred that a train following a locomotive has but little effect on the speed at which the locomotive sets up a condition of resonance, but, to some small extent, it will reduce the violence of this maximum state of oscillation. Hence, for calculating dynamic allowances for a railway bridge, to be on the safe side, it is advisable to neglect the train load altogether, and to assume that the dynamic loading which has to be added to statical loading of the locomotive plus train is that due to the locomotive only, running at the speed which oscillates the bridge to the maximum extent.

Effect of anti-phased hammer-blows on a double-track bridge

In a double-track bridge it has been tacitly assumed that the maximum state of oscillation which has to be considered in computing dynamic allowances is that produced when the hammer-blows of the locomotives on the two tracks are in step. This assumption demands some justification, since it is conceivable that a more serious state of oscillation might be developed when the locomotives on the two tracks have their hammer-blows opposed in phase. The cross-section of the bridge under these circumstances is deformed in the manner suggested by Fig. 64, one supporting truss being deflected upwards while the other is deflected downwards to the same extent. This will be termed the "anti-phased" mode of oscillation in contradistinction to the "in-phased" mode produced by the hammer-blows when they are in step.

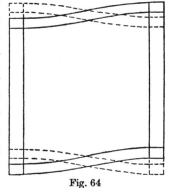

Fig. 64

When the frequency of the "anti-phased" hammer-blows coincides with the natural frequency of the "anti-phased" mode of vibration, oscillations of this character will be most pronounced, and a rough approximation to this frequency can be obtained by imagining the transverse beams to be pin-jointed to the main supporting trusses.

Suppose the distance between the centre lines of the trusses is $2a$ and the distance between the centre lines of the two tracks is $2b$, see Fig. 65.

Let M_D denote the total mass of the decking and tracks, and M_T the total mass of each truss.

For the "anti-phased" mode of vibration, the mass which has to be added to a truss to account for the inertia of the decking in its see-saw type of motion is $\dfrac{M_D}{6}$ approximately, and the fundamental frequency of the main trusses will be given by

$$2\pi n_0 = \frac{\pi^2}{l^2} \sqrt{\frac{EIl}{M_T + \frac{1}{6}M_D}},$$

whereas the frequency for the corresponding "in-phased" mode of vibration is given by

$$2\pi n_0 = \frac{\pi^2}{l^2} \sqrt{\frac{EIl}{M_T + \frac{1}{2}M_D}}.$$

If, in addition, the bridge carries two masses M, one on each track at the centre of the bridge (as indicated in Fig. 65), the corresponding formulae for the loaded frequencies are

$$2\pi n_1 = \frac{\pi^2}{l^2} \sqrt{\frac{EIl}{M_T + \frac{1}{6}M_D + \dfrac{2b^2}{a^2}M}} \quad \text{for the "anti-phased" motion,}$$

and $\quad 2\pi n_1 = \dfrac{\pi^2}{l^2} \sqrt{\dfrac{EIl}{M_T + \frac{1}{2}M_D + 2M}} \quad$ for the "in-phased" motion.

Thus it appears that both the unloaded and loaded fundamental natural frequencies for the "anti-phased" mode of vibration are likely to be somewhat higher than those of the corresponding "in-phased" motion, but on long-span bridges the difference will not be great, since in such bridges the main trusses contribute a relatively large proportion of the mass of the bridge. For a long-span bridge, if synchronous oscillations of the "in-phased" type are set up when the engine revolutions are N and the hammer-blows have an intensity $P_1 N^2$, then for synchronous oscillations of the "anti-phased" type the hammer-blows will have

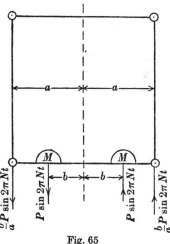

Fig. 65

the somewhat higher value $P_1 N'^2$ and on this basis of comparison, and assuming that the damping influences in these two cases are the same, the "anti-phased" oscillations will be the greater. It must, however, be noted that for opposing hammer-blows $\pm P \sin 2\pi Nt$ on the two tracks, the alternating forces acting on the main trusses are reduced to $\pm \dfrac{b}{a} P \sin 2\pi Nt.$

In any practical case $\dfrac{b}{a}$ can hardly exceed $\frac{1}{2}$ and, consequently, the fact that synchronous oscillations of the "anti-phased" type are associated with larger hammer-blows than those producing "in-phased" synchronous oscillations is more than compensated for by the fact that the direct action exerted on the main trusses by the "anti-phased" hammer-blows is much less effective.

Accordingly, for double-track long-span bridges, for the design of the main trusses, the greatest dynamic allowances are required when the hammer-blows of the locomotives are in step.

For medium-span bridges, synchronism with the "in-phased" mode of vibration can hardly be reached at practicable engine speeds and "anti-phased" resonance will be even less attainable.

For short-span bridges in which the maximum dynamic effect is always associated with the highest possible speeds, the reduction in the effectiveness of "anti-phased" hammer-blows removes the necessity of giving serious consideration to oscillations of this character.

It is only when the secondary dynamic stresses in transverse members due to the deformations indicated in Fig. 64 have to be calculated that "anti-phased" oscillations need be taken into account, and for this purpose the relative displacement of the main trusses can be estimated by assuming that they are subjected to moving hammer-blows of magnitude $\pm \dfrac{b}{a} P \sin 2\pi N t$, and that the natural frequencies of a truss loaded and unloaded are given approximately by the formulae stated above.

Apart from the stress in cross-beams due to "anti-phased" oscillations in the main trusses, the dynamic effects in the floor section of a bridge are free from complications. The members constituting the decking have, in general, such high natural frequencies that the operative forces they encounter may be regarded as practically non-dynamic in their effect. Consequently, dynamic allowances in these members can be taken into account by the simple process of superposing the operative forces on the statical loads to which these members are subjected and, when the state of oscillation in the main trusses has been determined, the computation of these operative forces presents no difficulty.

SYNOPSIS

For convenience of reference, the most important results arrived at in the foregoing Chapters are collected together and re-stated in this final summary.

NOTATION EMPLOYED

Bridges are idealized as girders of uniform mass and section freely supported at their extremities.

M_G is the total mass of the bridge;

l is the length of the span;

$m = \dfrac{M_G}{l}$ is the mass per unit length;

I is the relevant moment of inertia of the cross-section;

n_0 is the fundamental frequency of the unloaded bridge;

n_b is the damping coefficient;

$e^{-2\pi\frac{n_b}{n_0}}$ is the common ratio of successive residual oscillations for the unloaded bridge;

D_P is the central deflection due to a central statical load P.

A locomotive is idealized as a mass, spring supported on a single pair of wheels and axle.

M is the total mass of the locomotive (excluding tender);

M_S is the spring-borne mass;

M_U is the unsprung mass;

n_s is the frequency for free vertical oscillations of M_S on its springs;

n_d is the damping coefficient for spring movement;

F is the total frictional force resisting spring movement;

N denotes the revs. per sec. of the driving-wheels;

v is the speed of the locomotive;

$n = \dfrac{v}{2l}$;

$P \sin 2\pi N t$ is the hammer-blow at N revs. per sec.;

$P_1 N^2 = P$ is the intensity of the hammer-blow at N revs. per sec.

FREQUENCIES

The fundamental natural frequency n_0 of the unloaded bridge is given by

$$2\pi n_0 = \frac{\pi^2}{l^2}\sqrt{\frac{EIl}{M_G}},$$

where E, I, l and M_G are expressed in kinetic units.

If E is expressed in tons per square inch, M_G in tons, I and l in feet units,

$$n_0 = 10\cdot83\,\frac{\pi^2}{l^2}\sqrt{\frac{EIl}{M_G}}\text{ periods per second.}$$

The fundamental natural frequency of the bridge when carrying additional masses M_1, M_2, ..., concentrated at sections distant a_1, a_2, ... from an end of the span, is reduced to

$$n_0\sqrt{\frac{M_G}{M_G + 2\left(M_1\sin^2\dfrac{\pi a_1}{l} + M_2\sin^2\dfrac{\pi a_2}{l} + ...\right)}}.$$

DEFLECTIONS

For determining deflections, a force P concentrated at a section distant a from the end of the span can be replaced by a distributed load, the load per unit length being expressed in the form of an harmonic series

$$\frac{2P}{l}\left(\sin\frac{\pi a}{l}\sin\frac{\pi x}{l} + \sin\frac{2\pi a}{l}\sin\frac{2\pi x}{l} + \sin\frac{3\pi a}{l}\sin\frac{3\pi x}{l} + ...\right).$$

If P is a stationary force of constant magnitude, the deflection y at any section x is given by the formula

$$y = \frac{2Pl^3}{\pi^4 EI}\left(\sin\frac{\pi a}{l}\sin\frac{\pi x}{l} + \frac{1}{2^4}\sin\frac{2\pi a}{l}\sin\frac{2\pi x}{l} + \frac{1}{3^4}\sin\frac{3\pi a}{l}\sin\frac{3\pi x}{l} + ...\right).$$

On the assumption that contributions made to deflection by the higher harmonic components of the load distribution are negligible in comparison with those due to the primary component, the following results emerge.

D_P, the central deflection (measured in feet) due to a steady central force of P tons, is given by

$$D_P = \frac{Pg}{2\pi^2 n_0^2 M_G},$$

where M_G is measured in tons.

For forces P_1, P_2, P_3, ... tons, acting at sections distant a_1, a_2, a_3, ... from an end of the span, the central deflection is

$$\frac{\left(P_1 \sin \frac{\pi a_1}{l} + P_2 \sin \frac{\pi a_2}{l} + P_3 \sin \frac{\pi a_3}{l} + ...\right) g}{2\pi^2 n_0^2 M_G}\text{feet,}$$

where M_G is measured in tons.

For an alternating force $P \sin 2\pi Nt$ acting at a section distant a from an end of the span, the central deflection after the oscillation has become steady is

$$D_P \frac{\sin (2\pi Nt - \alpha) \sin \frac{\pi a}{l}}{\left\{\left(1 - \frac{N^2}{n_0^2}\right)^2 + \left(\frac{2Nn_b}{n_0^2}\right)^2\right\}^{\frac{1}{2}}},$$

where $\tan \alpha = \dfrac{2Nn_b}{n_0^2 - N^2}$, and D_P is the central deflection due to a steadily applied central force P.

For a force of constant magnitude P, moving at a uniform speed v along the span, the state of oscillation set up is the same as that due to a central alternating force of magnitude $P \sin 2\pi nt$, where $n = \dfrac{v}{2l}$.

For an alternating force of magnitude $P \sin 2\pi Nt$, moving at a uniform speed v along the span, the state of oscillation set up is the same as that due to the combined action of two central alternating forces of magnitude

$$\frac{P}{2} \cos 2\pi (N - n)t \quad \text{and} \quad -\frac{P}{2} \cos 2\pi (N + n)t.$$

The oscillations developed in a bridge due to the passage of a locomotive are greatest when the locomotive is not followed by a train. The increments of deflection thus deduced will provide dynamic allowances which err slightly on the side of safety when applied to cases where the locomotive is followed by a train. Allowances thus deduced are given for the four following cases:

Case (1)   Single-track bridge.
        Load, a single locomotive.

Case (2)   Single-track bridge.
        Load, double-heading, with two similar locomotives following one
           another at a distance d apart.

Case (3) Double-track bridge.

  Load, two similar locomotives, one on each track, arriving simultaneously at the centre of the span.

Case (4) Double-track bridge.

  Load, double-heading on each track, the four locomotives being similar and the locomotives in each pair being spaced a distance d apart.

Where two or more locomotives are on the bridge it is assumed that their hammer-blows are in step, although in the case of four locomotives the probability of this agreement is almost inconceivable.

CLASSIFICATION

Bridges are divided into three classes:

  (a) Long-span bridges of about 250 feet span and upwards.

  (b) Medium-span bridges, ranging from 50 feet to about 250 feet.

  (c) Short-span bridges.

Long-span bridges

A bridge is considered to be in this class if the oscillations to which it is subjected are never sufficiently vigorous to stimulate movements in the springs of a locomotive as it crosses the bridge.

If $\delta \sin 2\pi Nt$ defines the maximum vertical displacement given to the wheels of the locomotive, the springs will remain locked by friction provided that

$$\delta < \frac{Fg}{4\pi^2 N^2 M_S},$$

where F, the total force resisting spring friction, and M_S, the spring-borne mass of the locomotive, are both measured in tons and δ is measured in feet.

Frequencies at which the bridge oscillations are most pronounced

The maximum state of oscillation is developed when the hammer-blows synchronize with the fundamental frequency of the bridge at the instant when the locomotive is passing the centre, or, in the case of double-heading, when equally spaced on either side of the centre.

Consequently, in computing the maximum dynamical deflections in long-span bridges, N should be taken to have the following values:

Case (1)
$$N = n_0 \sqrt{\frac{M_G}{M_G + 2M}};$$

Case (2)
$$N = n_0 \sqrt{\frac{M_G}{M_G + 4M \cos^2 \frac{\pi d}{2l}}};$$

Case (3)
$$N = n_0 \sqrt{\frac{M_G}{M_G + 4M}};$$

Case (4)
$$N = n_0 \sqrt{\frac{M_G}{M_G + 8M \cos^2 \frac{\pi d}{2l}}}.$$

Dynamic increments of central deflection

Taking N to have the values stated above and $P \sin 2\pi Nt$ to be the corresponding hammer-blow, the following are the greatest dynamic increments of central deflection for which provision must be made:

Case (1)
$$\delta_0 = \frac{D_P}{2} \frac{N}{\sqrt{n^2 + n_b'^2}};$$

Case (2)
$$\delta_0 = D_P \cos \frac{\pi d}{2l} \frac{N}{\sqrt{n^2 + n_b'^2}};$$

Case (3)
$$\delta_0 = D_P \frac{N}{\sqrt{n^2 + n_b'^2}};$$

Case (4)
$$\delta_0 = 2 D_P \cos \frac{\pi d}{2l} \frac{N}{\sqrt{n^2 + n_b'^2}}.$$

In each case D_P is the steady central deflection due to a force P statically applied at the centre of the bridge.

For Case (1)
$$n_b' = n_b \frac{M_G}{M_G + 2M};$$

for Case (2)
$$n_b' = n_b \frac{M_G}{M_G + 4M \cos^2 \frac{\pi d}{2l}};$$

for Case (3)
$$n_b' = n_b \frac{M_G}{M_G + 4M};$$

for Case (4)
$$n_b' = n_b \frac{M_G}{M_G + 8M \cos^2 \frac{\pi d}{2l}}.$$

Live-load allowances to account for maximum dynamic bending-moments

The values of N and δ_0 to be employed are those, stated above, corresponding to the maximum states of oscillation:

Case (1) Live-load increment

$$=\left(M_G+\frac{\pi^2}{4}M\right)\delta_0 N^2 \text{ tons, uniformly distributed;}$$

Case (2) Live-load increment

$$=\left(M_G+\frac{\pi^2}{2}M\frac{l-d}{l}\cos\frac{\pi d}{2l}\right)\delta_0 N^2 \text{ tons, uniformly distributed;}$$

Case (3) Live-load increment

$$=\left(M_G+\frac{\pi^2}{2}M\right)\delta_0 N^2 \text{ tons, uniformly distributed;}$$

Case (4) Live-load increment

$$=\left(M_G+\pi^2 M\frac{l-d}{l}\cos\frac{\pi d}{2l}\right)\delta_0 N^2 \text{ tons, uniformly distributed;}$$

where δ_0 is measured in feet, M_G and M are measured in tons.

Allowance for dynamic shear

Taking N and δ_0 to have the values corresponding to the greatest state of oscillation, the maximum dynamic shearing-force at a section distant a from the further end of the span is given by the following expressions:

Case (1) Dynamic shearing-force

$$=\frac{\pi}{8}\left[-M_G\cos\frac{\pi a}{l}+M\frac{\pi a}{l}\sin\frac{\pi a}{l}\right]N^2\delta_0 \text{ tons;}$$

Case (2) Dynamic shearing-force

$$=\frac{\pi}{8}\left[-M_G\cos\frac{\pi a}{l}+M\frac{\pi a}{l}\sin\frac{\pi a}{l}+M\frac{\pi(a-d)}{l}\sin\frac{\pi(a-d)}{l}\right]N^2\delta_0 \text{ tons;}$$

Case (3) Dynamic shearing-force

$$=\frac{\pi}{8}\left[-M_G\cos\frac{\pi a}{l}+2M\frac{\pi a}{l}\sin\frac{\pi a}{l}\right]N^2\delta_0 \text{ tons;}$$

Case (4) Dynamic shearing-force

$$=\frac{\pi}{8}\left[-M_G\cos\frac{\pi a}{l}+2M\frac{\pi a}{l}\sin\frac{\pi a}{l}+2M\frac{\pi(a-d)}{l}\sin\frac{\pi(a-d)}{l}\right]N^2\delta_0 \text{ tons;}$$

where δ_0 is measured in feet, M_G and M are measured in tons.

MEDIUM-SPAN BRIDGES

For bridges whose oscillations are sufficiently active to induce spring movement, the greatest dynamical effect of a locomotive is determined by considering the locomotive to be stationary at the centre of the bridge skidding its wheels at a constant speed of N revs. per sec. For double-heading the skidding locomotives are spaced at equal distances on either side of the centre.

Values of N for which the bridge oscillations are greatest

N^2 is the greater of the two roots of the quadratic equation

$$N^4 - N^2 n_1{}^2 \left(1 + \frac{n_s{}^2}{n_2{}^2}\right) + n_1{}^2 n_s{}^2 = 0,$$

where $n_1{}^2$ and $n_2{}^2$ have the following values:

Case (1)
$$n_1{}^2 = n_0{}^2 \frac{M_G}{M_G + 2M_U}, \quad n_2{}^2 = n_0{}^2 \frac{M_G}{M_G + 2M};$$

Case (2)
$$n_1{}^2 = n_0{}^2 \frac{M_G}{M_G + 4M_U \cos^2 \frac{\pi d}{2l}}, \quad n_2{}^2 = n_0{}^2 \frac{M_G}{M_G + 4M \cos^2 \frac{\pi d}{2l}};$$

Case (3)
$$n_1{}^2 = n_0{}^2 \frac{M_G}{M_G + 4M_U}, \quad n_2{}^2 = n_0{}^2 \frac{M_G}{M_G + 4M};$$

Case (4)
$$n_1{}^2 = n_0{}^2 \frac{M_G}{M_G + 8M_U \cos^2 \frac{\pi d}{2l}}, \quad n_2{}^2 = n_0{}^2 \frac{M_G}{M_G + 8M \cos^2 \frac{\pi d}{2l}}.$$

In the formula which follows, the critical frequencies, obtained by solving the quadratic equation, are denoted by N_2. If the value of N_2 thus determined is above the range of practicable speeds, the highest permissible speed (say $N = 6$) should be taken, and this again will be denoted by N_2.

Dynamic increments of central deflection

Provided that N_2, as determined from the equation stated above, has a value which is within the range of practicable speeds, the corresponding maximum central dynamic deflection has the following values:

Case (1) $$\delta_0 = D_P \frac{n_0{}^2}{2N_2 n_b} \left[1 - \frac{4F}{\pi P} \frac{N_2{}^2}{N_2{}^2 - n_s{}^2}\right];$$

Case (2) $$\delta_0 = 2D_P \cos \frac{\pi d}{2l} \frac{n_0{}^2}{2N_2 n_b} \left[1 - \frac{4F}{\pi P} \frac{N_2{}^2}{N_2{}^2 - n_s{}^2}\right];$$

Case (3) $\delta_0 = 2D_P \dfrac{n_0{}^2}{2N_2 n_b}\left[1 - \dfrac{4F}{\pi P}\dfrac{N_2{}^2}{N_2{}^2 - n_s{}^2}\right];$

Case (4) $\delta_0 = 4D_P \cos\dfrac{\pi d}{2l}\dfrac{n_0{}^2}{2N_2 n_b}\left[1 - \dfrac{4F}{\pi P}\dfrac{N_2{}^2}{N_2{}^2 - n_s{}^2}\right];$

where F is the total frictional force resisting spring movement for a single locomotive, P is the hammer-blow at N_2 revs. per sec. for each locomotive, D_P is the steady central deflection due to a force P statically applied at the centre of the bridge.

If N_2, as determined by the equation

$$N^4 - N^2 n_1{}^2\left(1 + \frac{n_s{}^2}{n_2{}^2}\right) + n_1{}^2 n_s{}^2 = 0,$$

has a value which exceeds that attainable in practice, the oscillations will be greatest when N has the highest permissible value, say $N = 6$, and the corresponding maximum central dynamical deflections are as follows:

Case (1) $\delta_0 = D_P \dfrac{F}{\pi P}\dfrac{M_G}{M_S}\dfrac{n_0{}^2\{(N^2 - n_s{}^2)^2 + (2Nn_d)^2\}^{\frac{1}{2}}}{N^3 n_d},$

where n_d is the positive root of the quadratic equation

$$\left[\frac{2\pi^3 N^3 M_S D_P}{Fg}\right]^2 n_d{}^2 = [\Phi(N)]^2 + [\Psi'(N)]^2.$$

Case (2) $\delta_0 = D_P \dfrac{F}{\pi P \cos\dfrac{\pi d}{2l}}\dfrac{M_G}{M_S}\dfrac{n_0{}^2\{(N^2 - n_s{}^2)^2 + (2Nn_d)^2\}^{\frac{1}{2}}}{N^3 n_d},$

where n_d is the positive root of the quadratic equation

$$\left[\frac{4\pi^3 N^3 M_S D_P \cos^2\dfrac{\pi d}{2l}}{Fg}\right]^2 n_d{}^2 = [\Phi(N)]^2 + [\Psi(N)]^2.$$

Case (3) $\delta_0 = D_P \dfrac{F}{\pi P}\dfrac{M_G}{M_S}\dfrac{n_0{}^2\{(N^2 - n_s{}^2)^2 + (2Nn_d)^2\}^{\frac{1}{2}}}{N^3 n_d},$

where n_d is the positive root of the quadratic equation

$$\left[\frac{4\pi^3 N^3 M_S D_P}{Fg}\right]^2 n_d{}^2 = [\Phi(N)]^2 + [\Psi(N)]^2.$$

Case (4) $\delta_0 = D_P \dfrac{F}{\pi P \cos\dfrac{\pi d}{2l}}\dfrac{M_G}{M_S}\dfrac{n_0{}^2\{(N^2 - n_s{}^2)^2 + (2Nn_d)^2\}^{\frac{1}{2}}}{N^3 n_d},$

where n_d is the positive root of the quadratic equation

$$\left[\frac{8\pi^3 N^3 M_S D_P \cos^2 \frac{\pi d}{2l}}{Fg}\right]^2 n_d{}^2 = [\Phi(N)]^2 + [\Psi(N)]^2.$$

In each case F is the total force resisting spring movement (in tons) for a single locomotive, P is the hammer-blow at the highest permissible speed, say $N = 6$, measured in tons, D_P is the steady central deflection, measured in feet, due to a force P tons statically applied at the centre of the bridge,

$$\Phi(N) = \frac{N^4}{n_1{}^2} - N^2\left(1 + \frac{4n_b n_d}{n_0{}^2} + \frac{n_s{}^2}{n_2{}^2}\right) + n_s{}^2,$$

$$\Psi(N) = 2N\left(n_d\frac{n_2{}^2 - N^2}{n_2{}^2} + n_b\frac{n_s{}^2 - N^2}{n_0{}^2}\right),$$

the values of $n_1{}^2$ and $n_2{}^2$, for the four cases, being those stated on p. 199.

Live-load allowances to account for maximum dynamic bending-moments

The values of N_2 and δ_0 to be inserted in the following formulae are those corresponding to the maximum states of oscillation and are determined by the methods stated above.

Case (1) Live-load increment

$$= M_G\left[N_2{}^2 + \frac{\pi^2}{8}(n_0{}^2 - N_2{}^2)\right]\delta_0 \text{ tons, uniformly distributed;}$$

Case (2) Live-load increment

$$= M_G\left[N_2{}^2 + \frac{\pi^2}{8}\frac{l-d}{l}\frac{1}{\cos\frac{\pi d}{2l}}(n_0{}^2 - N_2{}^2)\right]\delta_0 \text{ tons, uniformly distributed;}$$

Case (3) Live-load increment

$$= M_G\left[N_2{}^2 + \frac{\pi^2}{8}(n_0{}^2 - N_2{}^2)\right]\delta_0 \text{ tons, uniformly distributed;}$$

Case (4) Live-load increment

$$= M_G\left[N_2{}^2 + \frac{\pi^2}{8}\frac{l-d}{l}\frac{1}{\cos\frac{\pi d}{2l}}(n_0{}^2 - N_2{}^2)\right]\delta_0 \text{ tons, uniformly distributed;}$$

where δ_0 is measured in feet and M_G in tons.

Maximum dynamic shear

The maximum dynamic shearing-force at a section distant a from the further end of the span is given by the expression $(S_1^2 + S_2^2)^{\frac{1}{2}}$ tons, where S_1 and S_2 are to be determined by the following expressions:

Case (1) $S_1 = \dfrac{\pi M_G \delta_0}{32}\left[-4N_2^2 \cos\dfrac{\pi a}{l} + \pi(n_0^2 - N_2^2)\right],$

$S_2 = \dfrac{\pi M_G \delta_0}{32} 2N_2 n_b\left(4\cos\dfrac{\pi a}{l} + \pi\right).$

For the region near the centre of the span the values for the resultant dynamic shear thus obtained must not be allowed to fall below the value $\dfrac{1}{2}\left(P_1 N_F^2 + \dfrac{M}{M_S} F\right)$ tons, where N_F is the frequency at which spring friction just breaks down and $P_1 N_F^2$ tons is the corresponding hammer-blow.

Case (2) $S_1 = \dfrac{\pi M_G \delta_0}{32}\left[-4N_2^2 \cos\dfrac{\pi a}{l} + \dfrac{\pi}{\cos\dfrac{\pi d}{2l}}(n_0^2 - N_2^2)\right],$

$S_2 = \dfrac{\pi M_G \delta_0}{32} 2N_2 n_b\left(4\cos\dfrac{\pi a}{l} + \dfrac{\pi}{\cos\dfrac{\pi d}{2l}}\right),$

the lower limit in this case being $P_1 N_F^2 + \dfrac{M}{M_S} F.$

Case (3) $S_1 = \dfrac{\pi M_G \delta_0}{32}\left[-4N_2^2 \cos\dfrac{\pi a}{l} + \pi(n_0^2 - N_2^2)\right],$

$S_2 = \dfrac{\pi M_G \delta_0}{32} 2N_2 n_b\left(4\cos\dfrac{\pi a}{l} + \pi\right),$

the lower limit in this case being $P_1 N_F^2 + \dfrac{M}{M_S} F.$

Case (4) $S_1 = \dfrac{\pi M_G \delta_0}{32}\left[-4N_2^2 \cos\dfrac{\pi a}{l} + \dfrac{\pi}{\cos\dfrac{\pi d}{2l}}(n_0^2 - N_2^2)\right],$

$S_2 = \dfrac{\pi M_G \delta_0}{32} 2N_2 n_b\left(4\cos\dfrac{\pi a}{l} + \dfrac{\pi}{\cos\dfrac{\pi d}{2l}}\right),$

the lower limit in this case being $2P_1 N_F^2 + \dfrac{2M}{M_S} F.$

In these formulae δ_0 is measured in feet and M_G in tons.

SHORT-SPAN BRIDGES

For bridges of 50 feet span and less, the maximum dynamic effects due to hammer-blows can be estimated by treating the hammer-blows as statical forces superposed upon the corresponding axle loads, the hammer-blows being computed for the highest speed permissible.

In all cases the allowances stated above for maximum dynamic and shearing-force are to be directly superposed upon the maximum bending-moments and shearing-forces due to the statical loads. Although the maximum dynamic and static effects at any given section may not always occur at precisely the same instant, the coincidence is sufficiently close to justify the process of superposition.